milk

The truth, the lies and the unbelievable
story of the original superfood

Matthew Evans

murdoch books

Sydney | London

Published in 2024 by Murdoch Books, an imprint of Allen & Unwin

Murdoch Books Australia
Cammeraygal Country
83 Alexander Street, Crows Nest NSW 2065
Phone: +61 (0)2 8425 0100
murdochbooks.com.au
info@murdochbooks.com.au

Murdoch Books UK
Ormond House, 26–27 Boswell Street, London WC1N 3JZ
Phone: +44 (0) 20 8785 5995
murdochbooks.co.uk
info@murdochbooks.co.uk

 A catalogue record for this book is available from the National Library of Australia

A catalogue record for this book is available from the British Library

ISBN 9 781 92261 686 9

Cover and text design by Daniel New
Typeset by Midland Typesetters
Printed and bound by CPI Group (UK) Ltd, Croydon CR0 4YY

We acknowledge that we meet and work on the traditional lands of the Cammeraygal people of the Eora Nation and the Wurundjeri people of the Kulin Nation, and we pay our respects to their elders past, present and future.

10 9 8 7 6 5 4 3 2

 MIX
Paper | Supporting
responsible forestry
FSC® C171272

Advance praise for m𝖎lk

'This is a rich dive into milk and the bounties it offers.
It'll make you wish you had Matthew Evans on tap
every morning – well, *Milk* is the next best thing!'
Dr Norman Swan

'A revelation. Educational, relevant and eye-opening.'
Nagi Maehashi, RecipeTin Eats

'One man's entertaining and deeply informative crusade into the human
obsession we call milk – and a vigorous argument for us to keep drinking
it, *Milk* vividly exposes all that is fascinating about one of the world's most
important foods with surprising facts and insights on every page.'
Dan Barber, chef, Blue Hill

'Unsettling and illuminating. You won't look at your bowl of cornflakes
in quite the same way after reading this riveting investigation. Milk is a
cultural, scientific and gastronomic feast. Evans at his best!'
Indira Naidoo

'The story of milk is so much more than milk in plastic ...
Matthew Evans, farmer and chef, does a brilliant job of taking you
step by step through the ages and stages of milk to present day,
milking milk, for all that it is and more. A brilliant read.'
Cyndi O'Meara, nutritionist

'... one of the best commentators on anything between the soil
and our stomachs.'
Jamie Blackett, *The Telegraph*

Also by Matthew Evans

• *Soil* • *The Commons* • *On Eating Meat* • *Not Just Jam*
• *Summer on Fat Pig Farm* • *Gourmet Farmer Goes Fishing*
• *Dirty Chef* • *The Gourmet Farmer Deli Book*
• *Winter on the Farm* • *The Real Food Companion*
• *Weekend Fare* • *Kitchen Basics*

To my mother, Barbara, whose milk made me who I am today.

And to Hedley, my son, whose love of warm milk in a bottle as
a toddler led to a much-loved 'hot milk dance' that I still (to his
embarrassment) mimic today when overwhelmed with joy.

Matthew Evans is an Australian chef, food critic and human biologist turned
farmer–food activist. He lives on a mixed farm that occupies a one-time apple
orchard in the Huon Valley outside Hobart, Tasmania. He is nationally best
known for his long-running SBS series *Gourmet Farmer* (six seasons), set on
Fat Pig Farm's 70 acres. He has authored or co-authored 15 books on food,
and is regularly interviewed on radio about all things to do with
farming, growing, soil and eating.

Contents

Fat of the Land

Thick, rich, custard-textured raw cream.

That's what I'm scooping out with my ladle. The milk churn sits on the bench. It holds 10 litres (or 2 gallons) of luscious, golden-tinged milk. I enticed it from Myrtle, my Jersey cow, at dawn yesterday and rested it in the fridge overnight. The oozy cream that has risen to the surface is easily plucked off the top with the lazy swirl of a ladle and the judicious use of a spatula. I hold in my hand the fat of the land.

Yesterday I used a litre of this milk to make yoghurt. I heated it in a pan until soft cappuccino-like bubbles appeared at the edges, and then I stirred it often as it sat, just below a simmer, for a quarter of an hour. Long, slow heat changes the milk, making my yoghurt thicker, richer, more densely textured, and hence fuller flavoured. I took the saucepan from the flame to cool it, to just above body temperature. This natural yoghurt is a living thing; the yeasts and bacteria essential to set and flavour it need warmth to thrive. I whisked in a tablespoon of a previous batch of yoghurt, inoculating the milk with the good bacteria found in natural curd, then poured it into a glass jar and insulated it in a small cooler box. The milk then sat on the bench for eight hours, after which it had set firm – though not as firm as it will become over the next week in

1

the fridge. When I dig my spoon into it, my yoghurt will weep a type of whey, greenish in hue.

Today, I want to use this yoghurt to culture an overabundance of cream. I warm the cream, too, like the previous day's milk, but only to the temperature of my blood. Body temperature is perfect for living things, and heating the cream further right now will change its flavour and its living flora in ways I don't desire. I add a tablespoon of yesterday's yoghurt to the cream, whisk briefly, then pour it into a jar to ferment for the next few hours. I'm making soured cream, or *crème fraîche* in French. It could find a use on tacos, in cake, over baked potatoes … but this batch I'll clot. Once soured, it'll be gently heated in a wide tray in a low oven until it sets and has a crust on top.

Clotted cream is a culinary thing of beauty; the combination of one of nature's gifts (cream), controlled rot (bacteria and yeasts) and human ingenuity (heat). Cracking through the caramelised skin to the dreamy sunset-yellow mousse below is one of the joys of milking a cow and being a cook. This masterful concoction will grace my porridge, my scones. I might even make a batch of ice cream from it.

Three days ago, this cream was grass. And that blows my mind.

In a day's time, the milk underneath the cream, which stills holds quite a bit of the fat thanks to my low-tech skimming, that milk will be feta cheese. A tablespoon of the whey drained from that cheese-making will be used to speed up the ferment of my sourdough bread. The rest of the whey will be used to make ricotta. Meanwhile, today's milk will be skimmed tomorrow to make butter, which will shame-lessly slather that bread. Welcome to the miracle of dairy.

This book was born out of my love of what comes from a cow. But it's also a celebration of all the milky goodness in the world. Goat's curd. Basque sheep milk cheese. Milkshakes. Breastmilk. Old-fashioned sherbets. Cultured butter. Cheesecake. Even almond milk, when it's made right.

When I say the word 'milk', you probably think first of cow's milk. This usually white liquid can appear deceptively plain and innocuous. It can come across as dull, because it's so ubiquitous; you can buy it everywhere, all the time, and it's often cheaper than bottled water. You can 'keep a cow in the cupboard' with powdered milk that lasts for years, or hoard a stash of UHT – the ultra-heat-treated milk that can be stored out of the fridge for months prior to opening.

While many of us only think about milk when we're pouring it on our cornflakes, it's also the stuff of international commodity markets, multinational processing companies, and aggressive anti-dairy campaigns. Milk was used by the temperance movement to prevent immoral and excessive behaviour, while glugging milk straight from the bottle, as we'll see later, has even been used as a symbol of anti-Semitism. Even human breastmilk – the original milk for us – has been fashionable and not, and is traded on the black market, the focus of body-builders and fetishists. How we see milk, in all its guises, has changed over the centuries, from hero to villain, lambasted or revered – driven by fashion, fanaticism and finances.

I know not everyone loves milk as I do. Some see the subjugation of a cow, any animal, as a form of modern slavery, a practice they suggest humanity will abhor in generations to come. Some can't drink milk. Won't drink milk. Don't drink milk, for physical or religious or philosophical reasons. And that's fine, as adults.

But many do drink it. It's believed three-quarters of the world's people get a substantial nutritional benefit from the milk of species not our own.[1] This can seem odd at a time when we're told that cows are causing climate change, that we don't need a diet containing animal products to have complete nutrition, and that plant-based eating has the lowest footprint on Mother Earth. With so many new 'milks' appearing on the market, my aim in this book is to talk about what milk is – to unpick what is good, and bad, and perhaps not yet widely known, about drinking the milk from humans and other animals, as well as the substitutes made from plants, so we can appreciate whether we should drink it at all.

As we'll see, milk's place in our history, in our culture and on our tables, has always been controversial.

The magic of milk

I'm a chef by trade, and a farmer and writer by desire. On our 70 acre (28 hectare) farm in southern Tasmania we grow 10,000 meals a year, from the vegetables to the pork, from the fruit to the beef. I also milk a cow. Alongside that, I spend my life working to improve the environment that we farm. Growing food, we think, shouldn't cost the Earth, and we're always trying to work out ways to nourish people while nourishing the landscape. Biodiversity, soil health, animal welfare – these aren't abstract topics on our farm. We live the reality of nurturing food from the land every single day. And yes, milking a cow is part of that.

For me, the journey into milk was kickstarted as a pre-teen. After a bushwalk near Bundanoon in Australia's south-east, we stopped for Devonshire tea after spying a hand-painted sign outside a farmhouse. While the woman of the house whipped up a quick batch of scones, she sent her husband out with us kids to milk the cow. Thinking about this now, it must have all been a show, because the cow was presumably milked at set times, not just when people visited. But I remember the milk, squirted into our faces by the farmer. I remember the flaky scones, topped with strawberry jam and whipped raw cream. I remember the smell of the cow, and such lovely milk. I can recall the smell of the dairy, too. All grass and hay and cow poo and musk. I remember being blown away that what embellished my scone had come from the udder of a benevolent bovine. It was this moment that made me one day want to milk my own.

It seems too incredible that you can conjure milk from a 500 kg (1100 lb) animal – one that could easily take you out with a well-placed kick – and that this milk could be curdled, spun, coaxed, solidified and become all those majestic culinary gems: butter, clotted cream, cloth-bound cheddar, panna cotta, paneer, Parmigiano Reggiano. It could be argued that the most concentrated expression of milk, combined with human gastronomic mastery, is in the set curds of cheese. The nutty aromatics and lightly firm texture of a Swiss Comté. The herbaceous aromatics of a Bonnet goat's cheddar from Dunlop in Scotland. My mate Nick Haddow's raw milk C2, a cooked curd cheese from Tasmania. The creamy, salty tang of Roquefort from the caves of Aveyron in France. I salivate just writing about them.

My first milk fixation – after I was weaned, of course – was cow's milk. The milk of my Celtic and Anglo forebears. It was the milk of an Australia in the 1970s, hampered somewhat by the free (but warm) milk given away at primary school to ensure we were all well fed. Prime Minister Gough Whitlam's ill-fated school milk program mirrored those in the UK and US, a well-meaning attempt to put milk, one of the most complete foods known to humanity, in the mouths of even the most vulnerable. When former British prime minister Margaret Thatcher removed their free milk program, the headlines read: THATCHER, THATCHER, MILK SNATCHER.

I remember the small glass bottles of unhomogenised milk, topped with cream and a foil cap, sweating outside the school tuckshop before being handed around at play lunch. We had to drink it, often slightly soured from its lack of refrigeration, its texture and temperature unappetising; thankfully a smuggled sachet of strawberry Quik (or Nesquik, as it's now known) helped get it down. And the payoff was that we could frisbee the foil lids around the schoolyard during play lunch.

For many, dairy products can hold both fond and foul memories. A favourite tuna mornay, the creamy steamy pasta topped with crisp breadcrumbs. A bottle of turned milk, poured into tea, its long past use-by date suddenly obvious from the clots that float to the top. Milk still warm and funky from the cows, drunk in a youth hostel in Switzerland, where I slept on the hay the cows would eat come winter. High-top apple pie served with an impossibly rich double cream. Cheese that smells like a teenager's socks, or Danish blue cheese so acidic it feels like your gums will recede. The perfect baked goat's milk ricotta in a hilltop town in Umbria. The rancid yak butter tea of a mountain village in Nepal. An oozy, melting, ripe brie that is too good to share. The perfect caffè latte.

Milk is an amazingly complex, versatile, nutritionally complete food that many cultures have venerated as a gift from the gods (the 'land of milk and honey').

I want people to marvel at milk. To cherish its role for us, from our first drink after birth, and for all mammals, and to understand what milk has done for us, and can continue to do for us, if only we value it more.

To do this, we'll delve down into the microscopic and gaze high up into the stars. We'll dip into the ocean's depths and explore the deserts. We'll look at religion, sexual politics, ultra-processed food, and human and planetary health. We'll confront birth, death, fetishes, and the supposed natural way of things. I am about to take you on a journey from the origins of life to the froth on your coffee while turning your understanding of nutrition on its head.

Come with me, as I try to unravel some of milk's mysteries and shine a light on its importance today, yesterday and tomorrow.

The Original Superfood

If a mammal doesn't get milk when it's first born, it usually dies. So, I'm worried.

I scoop Ailsa up in my arms and tuck her into my oilskin vest, next to my chest. Bedraggled and soaked through, Ailsa is a bony kid goat just a handful of days old and is quivering with the chill. It's 3°C, just above freezing, and she's looking hollowed out, weak, and on the verge of death. Her mum, Bianca, walked off while I watched, abandoning her daughter. Sometimes nature is cruel. I've seen mothers do this many times before, turning aside from their young just as Bianca has done. She's focusing on her other twin, because Ailsa doesn't have the strength to drink, and Bianca doesn't have the capacity to make her.

I'm no mother, but today that role falls to me. It becomes a lesson in the fragility of life and the vitalness of milk, a substance tweaked and honed through evolution over millions of years. An elixir so essential, and so much a part of human existence, that we go from revering it, to squandering it, to vilifying it on an almost daily basis. Today I will learn the hard way what it means to suckle young.

Before I carry Ailsa home and play mum, I've wiped her bum on the grass, removing lucid green poo – a blend of water and mucus.

Some still sticks to my hands. It smells vile. I've never seen, in 15 years of caring for animals, something coming out of a mammal's back end that is the colour of a light green jelly bean, and it scares me. When a kid (or a calf, or a lamb, or a human baby for that matter) is first born, the original poo is a form of dark faeces called meconium; the first passing of the bowels is stained almost black from the amniotic fluids of the womb. Colostrum, the early milk that a mammal drinks straight after birth, leads to bright orange poo in the first few days, and I've seen Ailsa pass some of that. But now her poo is Kermit green – translucent, tacky and weird.

I have little hope. I've nursed lambs on my chest before, rejected by their mothers and fading as I frantically tried to encourage a few drips of milk down their throats. I've lost piglets and calves who've looked much healthier than little Ailsa. I've used everything from milk-soaked cloths to buckets to eye-droppers to try to save our animals, attempting to do what nature has failed at, to keep its children alive. I know that what I do now will be the difference between Ailsa surviving or not – but her only hope is not me, really. It's milk.

I bring Ailsa to the house and towel her dry. In the freezer I keep colostrum, saved from the abundance that our dairy cows produce after giving birth. Colostrum, the rich yellow milk produced during the first few days of lactation, has more fat, protein and antibodies than later milk. It's like the milk of milk. Give a mammal colostrum in the first few hours of life and it can mean the difference between sickly and healthy. Between life and death. I know Ailsa got her mum's colostrum at birth, but I also know a small amount now will enrich any milk I manage to get into her.

I mix cow colostrum with goat milk, to make the concoction closer to what her mum would provide, and warm the mixture over a water bath, to her mum's body temperature. From her deep, ventral shudders and the way her head lolls, I know trying to get Ailsa to drink, to live, is a long shot. It's a very long shot if I can't get anything into her. And I can't.

Ailsa won't drink. Can't drink. She's lost her ability, or her will, to feed. I force her to try, holding the bottle vertically like her mother's

8

teat. I'm bending Ailsa's head back to reach the rubber nipple, and she suddenly finds some vigour – the vigour to fight back. She pulls away. She doesn't swallow. She lets out a faint, weak bleat. Again and again I try, the palm of my hand holding her lips tight to the teat. I stroke her throat gently with the fingers of my other hand to encourage the swallow reflex.

After about five minutes of this I'm about to give up. Then, a gulp. Then … nothing.

A single gulp, and at most about a tablespoon of milk has gone from the bottle. And only half of that into her system, I guess, because some milk dribbles from her mouth and across my fingers. Some has wet my jeans. We battle on in vain. She isn't swallowing and I can't get her to drink again.

She and I repeat this routine throughout the day, whenever I get a chance. Ailsa passes more mucus. She is hunched, hollow and listless. Any other kid her age would have leapt from her tub and gone exploring by now. But not Ailsa. She hasn't moved between each two-hourly feed. She is panting fast and shallow and her manner is forlorn, her ears drooping, her eyes dull.

Ailsa hangs on, barely, all day. She looks drier, less uncomfortable, but has only had about 50 ml – less than 3 tablespoons – to drink by 5 pm, judging by the way she swallowed at each feed. She's been quiet, apart from weak, plaintive bursts of annoyance when I've force-fed her.

With her head poking out from my vest, my warmth against hers, Ailsa and I sit for a while and rest. Then, suddenly, she bleats with gusto. And then she kicks. Even wrapped tightly against my chest I can feel a surge of strength. I pull her out and she looks glossier. Her eyes are bright, alert. Ailsa's ears are suddenly perky and her nose twitches as she lets rip another bleat. Her breathing is slower, deeper, her legs eager to run.

I step out to the top paddock. Ailsa calls loud and firm when she sees the flock. Bianca, her mother, bleats back and is instantly watchful. Popped on the ground, the previously incapacitated kid runs to her mother and starts to suck. Bianca stands proudly suckling

Ailsa while her twin brother looks on; Ailsa's tail twitching with joy, like a metronome on fast forward, as she drinks.

In hours, I've seen the transformation of inevitable death into hope. There must be dust in the air because my eyes water as Ailsa drinks. She slurps and cavorts as though today didn't happen. As if she'd never been at death's door.

After that one day, Ailsa thrived, without any input from me – showing the true miracle of milk. For all mammals, the difference between life and death lies, evolutionarily, in the ability of the young to suck and the mother to suckle. For a while I played mum, when nature faltered. But I'm no mother. I used nature – colostrum blended with milk – to save her. Yes, I did towel dry Ailsa. Yes, I warmed her up. But what brought her back to strength was the milk I managed to feed her, though it came almost too late.

It can be easy to forget the astounding life-giving power of milk, but for Ailsa, and for all mammals, milk is the original superfood.

If a mother elephant dies while the calf is still reliant on her milk, even the elephant's aunts can't save it. A giraffe needs to drink milk in its first few hours or it will fade and die. All mammals thrive on the milk that they were designed to have in the first few weeks, months, or even years of life. Humans are no exception.

When we're distracted by politics, or plant-based milks, or cow burps and the climate, it can be easy to forget that for most of human history, if *our* babies didn't get milk, they died too.

Take the Dublin Lying-In Hospital of the late 1700s and early 1800s. Of the babies born in this maternity hospital at that time, 99% died if they weren't fed by their mothers. 'Want of milk' was the stated cause of their demise.[1] Before we had infant formula, lack of milk for a baby was almost always a death sentence. Much later, at the start of the 20th century in Derby in the UK, the infant mortality rate (the number of infants who die before the age of 12 months) was three times higher in artificially fed babies – those fed by bottle – than breastfed babies.[2]

Even today, getting milk early, very early, can make all the difference to a child's chance of survival. A recent study from Ghana shows that a baby fed more than 24 hours after birth is twice as likely to die over one fed within the first hour.[3]

Milk is nutritional magic. It changes by the hour, by the day, by the mother, and certainly by the species. And milks aren't always interchangeable. As long ago as 1741, a physician named Hans Sloane believed that feeding newborn human babies with cow's milk was likely to be fatal. (Today we know that this was partly from the hygiene problems of the time, and partly because cow's milk isn't a complete food for a very young human baby.)

What's in milk is affected by everything – the individual mother, the individual baby. It changes by season, by diet, by social status, by the offspring's gender, by the stage of lactation and by the way it is handled. What an animal, including a human, gets from milk also changes as their gut matures.

Yes, all milk is truly amazing. Of course, it's vital early in life for all mammals, and keeps us alive in all kinds of ways. Yes, it has carbohydrate and fat and protein – all the macronutrients that we need to survive. It's probably why over 6 billion people on the planet use dairy as a nutritionally important component of their diet.[4]

But it also has properties that we are only just on the cusp of understanding. We now know that milk is living. It has microbes in it, establishing in the bellies of newborns a microbiome that is essential for proper digestion and more – all gifted from our mothers.

We also now know that milk contains things that affect our immunity, and our mental health, along with our ability to metabolise sugars and our ability to grow – but even more surprisingly, it also contains things that can change the expression of our genes. There are nano-sized components that can reach inside our brains and have an effect on the way we learn, think and remember. What's in milk, even cow's milk, can reach the very periphery of our bodies.

Milk makes us when we are young, and it can still be making us when we get old – even if we no longer drink it.

Starry, starry night

There's a brilliance to the night sky here in southern Tasmania. Well away from the humidity of the tropics, far from the pollution of our biggest cities and the all-around haziness of the northern hemisphere, I am witness to a multitude of stars that you can't see almost anywhere else. Even from bed I can spy the southern hemisphere's famed and most recognisable constellation in the heavens above us, the Southern Cross. Parked at the southernmost end of the Milky Way, the light band of sky smudged white by distant stars, the Southern Cross adorns our national flag as well as our cloudless evenings.

I go out on a bone-jarringly cold and clear winter's evening, admiring the Milky Way, the galaxy we call home, so named for its milk-spattered look from over 100 billion pin-pricks of light, each one a star. It's more than just the name of *our* galaxy that has my attention, though. It's the word 'galaxy' itself – which derives from the Greek *galaxias*, meaning milk. To the ancient Greeks, the Milky Way was *Galaxias Kyklos*, meaning 'milky circle'. Over time, 'galactic' became 'lactic', giving us the origin of our own milky terms *lactate*, *lactation* and *lactose* (even *lettuce*, because of the 'milk' that oozes from its stem when cut). The magnificence of the sparkling sky above is named for the humble white liquid that adorns the top shelf of my fridge. We even name a cloud formation, mammatus, after its apparent resemblance to milk's origin, the mammary glands.

Throughout history, milk has played a majestic role in all human lives, hence its immortalisation in the realm of the heavens. It's the first liquid to touch our lips after we're born. For aeons it was venerated, adored and admired, with attempts to replicate it outside the body going back as far as our recorded history. Gods such as the Egyptian Isis were shown suckling – in her case, nursing pharaohs. In the creation stories of Hinduism, part of the universe was originally seen as an ocean of milk, where gods and devils fought, and which the gods churned to create butter – creating order out of chaos (or just some nice ghee to put in your dhal).

The earliest evidence for humans drinking the milk of other species dates back about 8500 years,[5] though we were probably milking cows a couple of thousand years prior to that.

To those early farmers and graziers, milk was the original regenerating, sustainable superfood. To milk a sheep, it needs to give birth, and half the time the new lamb will be female, potentially growing up to also be milked. The male lambs became dinner a few months later, while the females promised a future full of dairy.

Milk was a mobile tap for high-quality protein, along with just about all the other nutrients that a human needs to survive, such as calcium, phosphorus and fats. What's more, this food can't go off if it's kept in the sheep, cow, donkey or buffalo, unlike those other high-quality proteins – meat and seafood – which would take processing to preserve them. Milk stored on-call in an animal is also more handy than beans and legumes, which need to be dried in order to be eaten out of season.

Once or twice a day, a complete food could be garnered fresh from the animal. What's more, not only did that animal carry the milk around for you until you were ready to milk her, but she could also walk to feed herself between camps or towns, and her protein couldn't be stolen by rodents. Imagine the labour saved, the nutrients upcycled, turning non-human forage, grass and trees into animal fodder, while being assured of a steady supply of a nearly complete food. The ancient Anglo–Saxons even called the entire month of May *Rimilcemona*, meaning 'the month of three milkings', when the longer northern hemisphere days coincided not only with the brightest moon but also the best grass, which meant more abundant and richer milk from their cows. May's full moon (or November's full moon in the southern hemisphere) is still sometimes known as the Milk Moon.

Milk is, simply, a superfood. The way mammals make it, what it does for their young, what it contains, and how it affects everything from gut health to neural pathways and the immune system, is utterly astonishing.

It's also a miracle in how it's made. A cow turns grass – essentially cellulose, a sugar we find indigestible – into milk in a matter of hours. Milk, which is high-quality protein, the most complex lipid (fat) in nature, has components that can enter the cerebral spinal fluid, the gel-like substance in our spines. For us, this is true not just of our own mother's milk, but cow's milk, too.

A human mother can transform her dinner of meat, vegetables, nuts, eggs and the like into the perfect safe food for her newborn, complete with a tailor-made set of antibodies triggered by the baby's mouth, smell and feel.

A piglet gets a meal of milk that contains thousands of species of bacteria, some passed from the mother's intestinal tract in a pathway yet to be identified by science, the milk altered by the sow's diet, but uncompromised by the fact that pigs can be copraphagic (poo eaters).

A blue whale exudes 200 kg (440 lb) of toothpaste-textured, high-fat milk into a slit in her side to suckle her young, which allows her calf to put on 100 kg (220 lb) of weight in a single day.

A baby goat goes from death's door to prancing around and leaping like a grass dolphin, simply through the power of milk.

Depending on the species, milk may be watery or rich, yellow or pink, thick or thin, salty, fishy tasting, sweet or none of those things. The processing, filtering, refining and delivery of this super-food from a lactating mother is a miracle of nature, an unfolding mystery to researchers, and way, way more complex than we previously imagined. If you thought you knew milk, think again.

Between our astral wonder and our deities, milk has been revered. And yet, today, it's easy to find milk demonised.

'In the US the FDA allows 750 million pus cells in every litre of milk,' claims Animal Liberation in their attempts to turn us all vegan. 'In Australia, New Zealand and Europe, regulators allow 400 million pus cells per litre,' they add.[6]

Unsurprisingly, as we'll see in later chapters, these statistics simply aren't true. Who knew that this remarkable liquid, the natural substance we drink more of than anything else around the globe, could be so polarising?

Or, perhaps we shouldn't be surprised. There's good evidence that ever since some of us first started consuming the milk of other mammals, others have considered it an abomination. This isn't something invented by the animal welfare movement of the 1970s. In the *Boke of Chyldren*, for instance, published in 1545, Thomas Phaire stated: 'If children be fed the milk of sheep, then their hair will be soft as that of a lamb, but if they be fed the milk of the goat, the hair will be coarse.'[7] Far more recently, the title of Mark Kurlansky's book *Milk! A 10,000-Year Food Fracas*, throws into focus the very long history of the whole controversy.

Before delving into the divisiveness of milk, let's take a collective deep breath and recognise the fundamentals.

Superfood basics

To survive, we humans need three major nutrients: protein, fats and sugar (carbohydrate). These so-called macronutrients are essential for all our major bodily functions – and milk has these in terrific amounts (and highly variable amounts, depending on the species). There's milk sugar, called lactose. There's protein, often in the form of casein, or other proteins, particularly whey proteins. And there's fat – or more accurately, fats – the amount of which can vary drastically, from below 1% to above 60%, depending on the animal the milk is from. The fats can fluctuate enormously in their exact composition, too. But enough of that for the moment; I'll devote a whole chapter to it later.

So far, so good. But milk contains much more than these big three dietary essentials. Besides lactose, it has other sugars that feed our own gut microbiome. It has complex proteins that affect our gut health. It has things called extracellular vesicles (EVs, and not the kind you'll be driving soon), which are tiny little fractions of stuff put out by other cells that play mightily complex roles in nutrition, despite their small size. There are also enzymes, antibodies, anti-oxidants and phytochemicals.

Unsurprisingly, the milk a mother makes for her infant is designed to match the exact needs of that infant – as a safe, perfect, singular food, its composition exactly as you'd engineer the food for a newborn, if you knew an animal from their atomic structure right through their cellular makeup to their place in the ecosystem. Milk is the evolutionary end point enabling mammals to feed their babies.

But what about drinking dairy after we've been weaned?

Domesticated animals used for dairy have been bred to produce way more milk than their calf, lamb or kid can drink. The modern dairy cow, for instance, can produce enough milk for up to 10 calves. While milk overproduction by dairy animals has consequences for those mothers, and for us (see the Cows With Guns chapter), animals that produce more milk than they need have helped nurse our babies, too. So much so that in the 1700s, orphanages kept goats or donkeys to feed the infants and prevent malnourishment.[8]

Dairy has sustained humans for generations, and is one of our most intriguing foods. It's not just the first food to be fully lab tested in the US; in most places it remains the most regulated food we put in our mouths. The results of our lactation, and those of other animals, also have deep cultural and gastronomic traditions. From the Moroccan *smen*, a very well cultured and aged butter that – legend has it – can be buried at your daughter's birth and dug up for her wedding, through Kenya's *mursik*, a thick sour milk drink stained grey from being stored in a charcoal-lined gourd, to the fermented yak butter and salt tea of the Himalayas that is given to all guests, milk's connections run deep in our humanity.

Globally, we're drinking more milk than ever. That's mostly in the Indian subcontinent and in China, and a lot has to do with an increasing population. In developed nations, however, milk consumption has plummeted. American milk consumption peaked in 1945 at a whopping 170 litres (45 gallons) per person per year, and dropped to just over 60 litres (16 gallons) by 2021. In Finland, milk consumption fell from about 350 litres (77 gallons) per year in 1950 to half that by the end of the 1950s, to half that again – about 85 litres (19 gallons) by the end of the 1970s.[9]

In the UK, consumption dropped from 140 litres (31 gallons) per person in 1974 to just 70 litres (15 gallons) in 2018.

Australia has fared a bit differently. According to Dairy Australia, milk consumption peaked in 2012/13 at over 106 litres (23 gallons) per year, and fell to 93 litres (20 gallons) by 2022.

Everywhere you look in developed nations, milk consumption has declined, while developing nations are tending to drink more of it. In richer countries, Gen Z – those born between 1997 and 2012 – are buying about 20% less milk than previous generations[10] – possibly due to the rise of plant milks, lactose intolerance, or animal welfare and climate change issues.

Or is it that they are just buying more junk? The great decline in milk consumption wasn't originally driven by animal or environmental concerns, but rather by people avoiding the fat in whole milk and consuming soft drinks instead – helped along by folks such as Fred Stare, who for a while as chair of nutrition at Harvard Medical School took secret funding from Coca-Cola in the 1960s. Stare encouraged people to avoid saturated fat in their diets (including dairy fats), while also writing letters in which he denied that Coke caused cavities. He also promoted 'soda, ice cream, or a Coke' as an appropriate mid-afternoon snack for teenagers.[11]

In the US, where good figures have been collated, the drop in milk usage coincided with a massive uptick in soft drink consumption in the 1980s and 1990s, followed by a later rise in waters and juices. Now it's the turn of plant milks,[12] with fake milk sales expected to grow 9% a year to 2027 in the US alone. Albeit from a low base.

To put this in perspective, in 2022 in the US, sales of alternative milks totalled $2.4 billion, compared with almost $15.7 billion for dairy milks. Because plant milks are more expensive, this dollar value skews the narrative, so volumes are harder to source. In the UK, alternative milks make up about 4.6% of the market, but are growing fast.[13]

If we really want to talk milk, however, then we need to talk about where milk comes from. To do that, there's no avoiding talking about the equipment that makes milk, and then allows it to enter the world … so there'll be talk of teats, nipples, breasts, mammary glands – terms that some may find a bit ick or salacious, but are natural to farmers, midwives and scientists.

For some, milk is the perfect food, to others it's the devil's drink … so let's start by taking a closer look at one of the most maligned mammals alive.

The Devil's Drink

I've been peering into the pouch of a devil. A real-life devil. A Tasmanian devil boasting 42 razor-sharp teeth[1] that will replenish themselves as they're worn down over her lifetime. Teeth designed for tearing apart flesh and grinding skulls. She has the fiendish, bone-chilling cry of something possessed, can smell the scent of her prey from a kilometre away, and puts out a dense, pungent stench to ward off threats. Tensed, balled up and made of muscle and might, if we lost our grip on her, we'd lose a finger. Or a hand.

I'm here because I'm more interested in her teats than her teeth. Milk brought me here, to risk the wrath of a devil, because what comes from the impossibly small teats of a Tasmanian devil has astounding properties that we are only just beginning to comprehend.

The milk with which a devil suckles her young is life-giving, but also a killer. It is known to kill golden staph (*Staphylococcus aureus*) – the world's worst superbug, an antibiotic-resistant bacterium that plagues our hospitals. Her milk can fight yeast infections such as candida. A devil's milk is also believed to kill vancomycin-resistant enterococcus, another of the dastardly microbes that are increasingly hard to treat with traditional antibiotics.[2]

I'm out here on a dawn raid peering into a devil's pouch because I want to understand milk in all its complexity. Far from the stuff on a supermarket shelf, far from the liquid that sustained our son when he was first born, milk is way more wondrous than we've ever imagined. To understand milk, we need to see it where it's naturally made – from the Southern Ocean to the hills of Tibet, from the massive indoor cow's milk dairies of Saudi Arabia, to a moss-fringed gully in southern Tasmania.

So, I was looking at devils because you have to know the animal, at some level, to understand their milk.

And to be honest, to understand milk, it would be good to have a firm idea of what it actually is. The problem is, we don't even really have a definition for milk that experts in the field can agree on. This evolutionary masterpiece may come from a nipple. Or not. It may be white. Or not. There's 'milk' from penguins, cockroaches, spiders, soybeans and rice. But mostly it's made by mammals, who take their name from the mammary gland that produces the milk, along with the subfamily of mammals, the marsupials, whose young are suckled in pouches.

Devils are the iconic Tasmanian native marsupial. After the extinction of another marsupial carnivore, the thylacine (Tasmanian tiger) in the mid-20th century, Tasmanian devils became the largest apex predator on our land. I've long been fascinated by our nocturnal visitors; they wake me from my bed at home. I hear their throaty, rasping, high-pitched growls carrying across the valley and into my bedroom as they fight over carrion. Relative to their body size, their compact jaws have the most crushing force of any animal,[3] able to chomp through bone (553 newtons,[4] if that's the kind of thing you like to know). They're the world's largest marsupial carnivore, whose Latin name, *Sarcophilus* (creepily close to the death bed sarcophagus we know from Egypt), means 'meat loving'.

Once common here, the Tasmanian devil is now considered endangered, plagued by a facial tumour that reduced their population by nearly 90% over 10 years in the early 2000s. I went trapping with Elise Ringwaldt from the University of Tasmania, who is helping to understand their plight. Today, along with other health measures, we want to see how many babies each female devil is suckling.

I'm fascinated by devils for lots of reasons, not least the amazing qualities of their milk. I'm also intrigued by their method of raising young, which, as with all mammals – including us – is intrinsically tied to their milk. A mother devil mates in early autumn with multiple partners, with copulation taking up to five days with each partner. In early April, after just three weeks pregnancy, she gives birth, releasing from one of her three vaginas up to 30 young, each about the size of a rice grain and weighing about 0.2 grams – so small that five of them weigh as much as a paperclip.[5] These babies, essentially little more than foetuses, must all frantically climb up her fur, through all the incumbent hazards of the outside world, and into her rear-facing pouch, where they will find only four teats. In an example of Darwinian theory in real time, the 30 barely formed joeys compete for the tiniest of nipples, and only the fastest survive. Over the next few months, those that won that race never leave the teat, clinging on like pegs as their mother fights and feeds at night, and hides in her den during the day. The milk they drink could well be death to golden staph, but it's life to them. Devil milk's ability to kill pathogens could well be to do with the joeys' early exposure to all kinds of microbes, some of them deadly.

Getting to the teat is paramount in a more basic sense, too. Elise points out that a devil's teat is like a substitute umbilical cord. All the siblings that don't make it to a teat die of starvation. There is nothing we – a supposedly clever species of bipeds that has travelled into space, and worked out how to split the atom – can do to save those babies if they don't get mum's milk.

By six weeks, the joeys are jelly bean–sized. They'll stay attached to the teats for 100 days. During that time, their transformation from a rice grain–sized baby devil into a thriving adolescent is entirely the result of devil milk. By the time they unlatch from their mum's teats, devil babies will have hair, and teeth, and a robust immune system. They'll also be 1000 times heavier than they were at birth, tipping the scales at 200 grams (7 oz).

Milk makes devils. It provides all of the nutrients needed to produce a mini-devil, complete with claws, eyes that can see in the dark, a biofluorescent coat that glows blue in the dark (yes, seriously) – and many other things we can't yet measure. What we do know is that all marsupial milk is incredible, given how the babies are born so little formed.

Remarkably, despite the tiny size of their joeys, Tasmanian devils don't give birth to the smallest babies of any mammal. That honour goes to the honey possum, whose babies are a miniscule 0.004 grams each at birth,[6] meaning it takes 15 of them to make up the size of a grain of rice. Again, they become a possum, a grown possum, through milk and nothing more.

How is this possible? What makes milk so good at its job? For that, we'll need to look beyond the usual. Beyond cows and goats and humans. But we'll also delve deep into the milks we *do* drink, from breast to camel to almond and oat. I'll explore why we have such a deep connection to milk, both emotional and cultural. To me, milk is a substance of precious beauty, whose mysteries we are only just beginning to unravel, and whose attributes we've often taken for granted.

For a start, we used to think milk was sterile when it arrived from a healthy mother. We now know it is laced with beneficial bacteria that help inoculate the newborn baby's immune system. Milk also has things in it that aren't digestible by the babies it's designed to feed, and now we are starting to understand why. What's in a mother's milk is affected by her mood, her diet, and her own gut health, and is exquisitely tailored to her baby's age, immune status and more.

And the milk produced during the later stages of breastfeeding is nothing compared to that first day's miracle: colostrum.

Colostrum: Making Milk Look Like a Try-Hard

Most mammals produce colostrum, a very early milk – the very *first* milk – which we met with Ailsa earlier. This densely nutritious, often brightly coloured concoction is designed to nourish the baby in more ways than one. Human babies drink less than a teaspoon at their first feed, often just a fraction of that, and they might only get up to 40–50 ml (1½ fl oz) of colostrum in the first day. A lactation consultant I spoke to said some mums can only express half a millilitre or a single millilitre at a time. It's a crazy small amount, yet that will be all the baby needs. When I say needs, however, the baby really, *really* needs it. Colostrum has health-giving properties that can last a lifetime – but babies need to get it early. Super concentrated, it is the bridge between the placenta and the teat, its role being to confer as much immunity to the baby as possible after the infant's passage from the safety of the womb to the challenging outside world. Colostrum is less a food than a vaccine.

Colostrum changes depending on the species, and it's made differently to milk, even though it comes through the same glands. In pretty much all the animals humans have looked at, colostrum is formed when the mammary gland is more open to the bloodstream, so it can take up more nutrients and bigger cells. That's why, if you've ever tasted it, it can be salty. So salty that a baby is usually not impressed by it even a week after they're born. From the baby's perspective, those early hours are vital, too; their gut is more porous for that first day, meaning they can take up these bigger molecules.

Human colostrum is deep yellow in colour, like the rich custard under a crème brûlée made with golden free-range eggs. And, like custard, colostrum is far fattier, and has far more protein, than later milk. There's a tiny fraction of milk sugar, too, but the important bit is protein. That's the bit that has antibodies and other good things that allow a vulnerable newborn with an immature immune system to be thrust into the teeming world of microbes in which we live. Colostrum is packed with a microcosm of the mother's acquired immune responses, in digestible and usable form.

The bigger molecules are there for a reason. Some can directly enter the infant's cerebrospinal fluid[1] – the gel-like substance that runs through the centre of our backbones to our brains, and is deeply involved in our nervous system. Colostrum is believed to help develop the brain and nervous system with proteins such as lacto-ferrin, and the immune system with things such as lysozyme, which I'll talk more about later.

Astoundingly, there are over 1000 different proteins in human colostrum. And about 40% of that protein (about double that of breastmilk, and 10-fold more than in cow's milk) is a protein called alpha-lactalbumin.[2] Nicknamed HAMLET ('Human Alpha-lactalbumin Made LEthal to Tumour cells'), it is a cancer-fighting agent that is being tested as an anti-cancer drug. We'll come back to HAMLET later, but for now, just know it's a cracker of an ingredient to give to a child.

Recent research also shows how intimately a mum's body influences her developing baby. It's been shown, in animal models at least, that maternal cells (from the mother's body) are found in the brains of infants. We don't know why, but maternal stem cells can make their way through the very challenging acidity of the baby's gut, into the bloodstream, then into the child's central nervous system, the baby's brain and spine.[3] Once there, being stem cells, they can actually form into more specialised body cells in the newborn – including glial and neuronal cells, which can transmit and regulate signals in the nervous system. A mother's stem cells become part of the brain of her baby. Like, crazy, yeah?

Colostrum is a powerhouse of goodness, packed with so many wondrous, seemingly magical things, including all kinds of growth factors, immunoglobulins, insulin and more. There are chemicals in human colostrum that help mature the infant's gut cell walls, and also repair them. They exist in milk, but there's 2000 times more in colostrum than in the later milk.[4]

Cow's colostrum is also deep yellow, but richer than a human's – high in protein and fat, and lower in sugar. Because a dairy cow produces way more colostrum than her calf needs, people have often used it in cooking, though its salty tang can be a challenge. In Britain it has been used to make a dessert called *beesting* (named after a German cake whose baker was apparently stung by a bee). There's the Swedish *kalvdans* (literally meaning 'calf dance'), made by gently cooking colostrum with sugar and letting the proteins set, kind of like a baked custard. And there's also the Indian *kharvas*, where colostrum is mixed with milk, jaggery (palm sugar) and cardamom, and then steamed.

Dried cow's colostrum is available in chemists. It's been shown to help treat diarrhoea, and some people use it in the hope that it will boost their immune system. A friend was recommended by her GP to take it while they were using anti-malarial medication, to help reboot their gut microbiome. Powdered colostrum is highly processed and sterilised, and free from pathogens and beneficial bacteria alike, but still contains some bioactive molecules. It is worth wondering, however: given that human maternal stem cells can enter a baby's brain, can cow stem cells enter adult human brains? It seems unlikely, because our gut has closed down some of its highly porous nature the day after we're born, but as this research is less than a decade old, such questions remain unanswered. And the cow colostrum you buy has already been pasteurised, killing any cells present.

Colostrum, in particular human colostrum, is relatively low in sugars, most notably lactose, which is the perfect fit for a newborn. Later milk contains more lactose, and that's right for that time of life, too.

While just about all of us can consume lactose when we're little, not all of us can do so with impunity when we grow up. It's often this milk sugar that causes some adults to steer away from dairy. And it all comes down to ancestry, and a mutant gene or two that set at least some of humanity on a lactose-loving course.

Mutant Milk Drinkers

Cari loved to drink milk, even as an 11-year-old. She drank it greedily and noisily, often spilling it as she supped. Bluey adores milk too. Both our farm dogs have cherished the flavour of milk. Many grown mammals do, including pigs and cats. It seems lots of mammals are hardwired to enjoy the flavour, sweetness and creaminess of milk, probably as a survival strategy from when they're first born. Cats so love the taste of milk that we often think they *should* be drinking it – the saucer of milk put out for our moggies a cliché mirrored in reality. This isn't something humans just foist on them. Even feral cats will seek milk out; they have been seen stealing milk from lactating elephant seals on Mexico's Isla Guadalupe, sucking at the seals' teats.[1]

Loving the taste of milk is probably an evolutionary advantage for mammals, and some of us never lose the attraction. Perhaps it's because of protein fragments called beta-casomorphins in milk, things that can pass from our gut to our brain and release the happy hormone serotonin.[2]

But liking the taste doesn't mean we should drink milk once we've been weaned off our mothers. Cats, for instance, probably shouldn't lick too much cream because – believe it or not – they're lactose intolerant. For humans, well some humans at least, it's a different story.

The thing is, all adult cats lack what some grown-up humans have: the ability to produce lactase – the enzyme that allows us to digest the sugars in milk, long after we've been weaned by our mothers. I say 'us', because although I'm in the minority, I'm one of those with the gene that allows my body to produce this particular enzyme.

Lactose is the most abundant sugar in milk, and without the enzyme lactase, a person is lactose intolerant. Their body can't break down lactose into a digestible form, and consuming too much milk compromises their gut. A cat, for instance, can only cope with about 6 g of lactose at a time, so to avoid gut irritation, they shouldn't drink more than 130 ml (half a cup) of milk a day. More sensitive cats, it's recommended, shouldn't drink more than 85 ml (a third of a cup) of milk in a day.[3] Remember that when you put the saucer out tonight.

All mammals, with very few exceptions (including some very rare cases in humans), are born with the ability to digest lactose. We produce the lactase we need in order to digest our first milk, allowing us to break milk sugar down into glucose and galactose and reap the benefit. Glucose is eminently digestible, being the basic sugar unit used in our blood; galactose is also an abundant sugar in our diets, easily converted into glucose by our livers.

After the age of about five years, only about a third of humans are able to produce lactase into adulthood.

Those of us who don't produce lactase, 65% of the global adult population, can't readily break lactose down into glucose and galactose, so that fraction of milk just passes through the gut mostly unchanged, causing irritation – bloating, gas, diarrhoea, sometimes constipation – on the way through. In other words, it's not nice.

The reason some of us can consume lots of lactose throughout life is because we're the descendants of mutants. Some time ago

– probably about 4000 or more years ago – there was a mistake in the human genome that changed the way people digested milk. Some of the earliest evidence of the gene's presence in Europe suggests there could have been what is called 'lactase persistence' in what is now modern-day Spain, though it's more likely to have originated in Sweden or Germany.[4]

Dairy regions, those that milked other mammals for human consumption, typically became lactose tolerant. At least five different gene mutations have happened around the globe, single substitutions in our DNA near where our lactase gene sits.[5] This glitch happened in the Middle East and multiple times in sub-Saharan Africa, as well as in Europe.[6] Because of migration, the genetic variations have been taken up in many places. While Europe is almost entirely lactose tolerant, and Pakistan is hugely lactose tolerant (particularly in the south), Africa is patchy.[7]

So, I'm puzzled by Genghis Khan. Well, not by this Mongolian warlord as much as by his armies. In the 1200s, the Mongols conquered an area of 31 million square kilometres – land roughly the size of the continent of Africa – riding on horseback from their base north of China. They controlled the largest empire on a single landmass in history. The empire ranged from modern-day Korea all the way to Istanbul, from India in the south to Novgorod near St Petersburg in the north. It took a lot of human power to fight these battles, with virtually all of them taking place a long way from the soldiers' homes. The singular, densely nutritious food the Mongols used to fuel this conquest was milk. Horse milk, garnered by milking their mares up to five times a day.

We've seen milk is nutritious, so the food source sounds plausible. But the Mongols lacked one thing, and it certainly wasn't the ambition to conquer, or bravery or resilience. What adult Mongols still lack today is lactase. And yet, mare's milk was their main source of travelling nutrition.

Not all dairying cultures developed the ability to digest milk. For instance, it's easy to understand how the indigenous people of the Americas didn't develop a mutation to allow them to produce lactase as adults, because milking animals wasn't a part of the culture there until European colonisation. Similarly, where there were no lactating animals producing milk of large volume – and hence no dairy culture existed, such as in Australia – there was also no lactose tolerance gene. Ditto in most of South-East Asia.

Mature Mongolians, however, despite milking animals for millennia, and despite milking multiple species, from horses and reindeers to cows and goats, lack the ability to digest milk in its unprocessed form – the form it comes direct from the animal.[8]

This anomaly points to one of the puzzles of lactose tolerance: the genetic variation only came about around 4000–5000 years *after* we starting milking species other than our own.[9] It seems a dumb idea to intentionally milk a cow or a horse to get a product you can't fully digest.

It's therefore quite intriguing. What happened prior to this mutation for all those dairying nations? And what happens – even today – to the Mongolians who not only built an empire thanks to dairy's nutrient density, but still spend a large amount of their time and effort breeding dairy animals, finding them fodder and enticing out their milk? Well, it turns out that – just like cats – a small amount of milk is usually okay for any human to drink. And most milk wasn't drunk as a fresh liquid anyway. Because there was no refrigeration, most milk was preserved in some way.

The bacteria and yeasts that inveigle milk when it is cultured thrive there because they, like the Mongols, find it nutrient dense. But unlike Mongolian dairy herders, these microbes just love lactose. *Lactobacillus*, which you may have seen on your pot of 'live yoghurt', is one family of lactose-eating bacteria that helps give yoghurt its trademark flavour and texture. These bacteria, and other culturing agents, consume lactose and convert it into lactic acid – something we *can* digest. And, what's more, these same bacteria are useful for gut health in the process. The microbes actually help food taste better, *and* help us digest and extract nutrients. Lactobacillus bugs also produce a whole heap of other beneficial compounds such as organic acids, exopolysaccharides and antimicrobial compounds that are not only pretty handy in helping make dairy more digestible, but are actually great for us more generally.

The Mongols' strategy is human-induced fermenting. Mongolians actually worked out how to dry milk out into a rubbery form, by culturing it, draining it through cloth, then sun-drying it in chunks. From that time hence, they could carry a lightweight version of their superfood across the Steppes and into battle. A little rehydrated mare's milk – condensed protein, fat and fermented milk sugars – in a chewable block that you could suck on even while riding your horse.

Apparently modern versions taste pretty bad, but if you were relying on human labour to conquer a continent, some fermented, rubbery dried milk could well be just the ticket.

Many of the things we still make from milk, such as yoghurt, kefir, cheese and the like, are all cultured (i.e. fermented) before we eat them. In the modern world, we can also add lactase to fresh milk, and create a non-soured drink where all those pesky milk sugars are pre-digested. Some lactose-intolerant people even swallow lactase pills to allow them to drink fresh milk – such is the desire to enjoy it in its pure state, or to be part of the culture of milk-drinking nations.

We also know that the microbiome of an individual, the trillions of bacterial cells in the human body – some of which are lactobacillus – can also help in the digestion of lactose. It's been suggested that because pre-modern humans had more diverse gut flora, those bacteria could break down lactose more readily.

While we humans are the only species to milk other animals for our own consumption on any great scale, we're not the only ones to have developed a tolerance for lactose later in life.

Curiously enough, in the same way that modern dogs can still thrive if they consume way more carbohydrate in their diet than their wolf ancestors (thanks to what we have fed them over thousands of years), many dogs are also able to produce lactase as adults. It seems that our domestication of them has altered their canine genetic makeup, too – so they, like us, have also mutated to develop a lactase persistence gene. Evidence shows it happened at about the same time as humans mutated. So, we're not the only known species that can digest milk post-weaning – though, just like us, not all dogs have the gene, and their individual ability to digest lactose can vary.

Given that all humans, with or without lactase, can digest some fresh milk and most fermented milk, why did the lactase mutation persist in our DNA? Science is still puzzled by lactase persistence in humans. We obviously don't actually need milk post-weaning, as shown by the majority of the world's population. Was it really such a boon to human fertility that the gene spread throughout populations where it occurred?

Lactase persistence gave an advantage of up to 4–5% per generation, according to the geneticists, which apparently is 'one of the largest selection pressures observed for any gene in recent human evolution'.[10] It's about on par with increased genetic resistance to one of humanity's modern plagues, malaria.

Just what this benefit actually *was* is still being debated. Did it help with reproduction, with disease, with childbirth? Did it allow bigger communities, hence more babies, than where the gene didn't arrive? Was the presence of a dairy animal an insurance policy against the death of the father, at a time when men were more likely to be killed at war, or the death of the mother, when women frequently died during childbirth? Or is it insurance against famine more generally? The Dutch became the tallest people in the world, overtaking the Americans sometime last century, thanks, in part, to their large intake of dairy. On average, they grew a massive 20 cm (8 inches) in 150 years.[11]

Darwinian theory suggests that lactose tolerance must have had an impact on childbirth and mortality rates to have spread so far, so fast, and so successfully. That said, the Mongols certainly proved that we don't need gene mutations to be able to utilise milk's nutrient density in other ways.

We may never really know why a third of humanity has inherited one of the lactase persistence genes. The fact that all five of the gene mutations appeared independently of each other, and then lingered and eventually dominated the genome where they occurred, means it has to have been of great value.

What we do know is that milk isn't essential for us as adults. Plenty of cultures have done well without it. In Australia we have the oldest continuously living culture, at about 60,000 years and counting, which survived without Aboriginal people consuming a skerrick of dairy except from their mothers or aunts in the first few years after birth. But the reality is that access to milk was, and perhaps still is, beneficial because it's a one-stop shop for so many nutrients that would be otherwise hard to source.

Lactose isn't the only component in milk that we may have trouble digesting, however. There are also proteins, and these vary widely in their amount and composition, depending on the animal that is milked. Because some proteins are harder to digest, even those of us who can digest lactose and aren't allergic to these proteins may have experienced some strange symptoms associated with eating dairy. One side effect can be 'cheese dreams' – the broken sleep of the rabid cheese fan.

But before we get down into such details, it's important to recognise that milk isn't just sustenance, it's an intrinsic part of human society. In the next chapter, we'll look at milk within the broader culture, to appreciate how even the virtuousness of breastfeeding has been used as a political and moral tool – and to understand why creepy pictures of a fully grown saint copping Mary's breastmilk in the eye started to appear in churches nearly a millennium ago.

Madonna's Breast

In many societies, part of the problem with talking about – and showing – lactation is that the female breast is a secondary sexual organ. Humans are the only mammal we know of in which the female breasts enlarge at puberty and stay enlarged throughout adult life, regardless of reproduction or lactation.

Throughout the ages, some women have decided not to breastfeed their babies because of what it might do to their breasts. As Clement of Alexandria stated around AD 200, after childbirth 'the breasts, which till then looked straight towards the husband, now bend down towards the child.'[1] Only recently, as a friend of mine attempted to nurse her newborn in a café, she was told by a stranger (of course it was a man), 'That's disgusting, put it away.' Things haven't changed much in the last 2000 years, it seems. The whole space is fraught. From sexual politics to actual politics, milk gets caught up in it all.

Milk's role in society is more than biological. It's also the stuff of myth and privilege, and an almost mystical power. Over 2500 years ago, near where the majestic city of Rome now stands, baby twin brothers were once suckled by a wolf. The newborns, named Romulus and Remus, had been abandoned by the side of the river Tiber, left to starve to death by their great uncle, King Amulius, who had just deposed their father to take the throne. The baby boys, however, were befriended by a wild wolf mother, who nurtured and suckled them until they were rescued by a local shepherd. Years later, a fully grown Romulus overthrew Amulius to reclaim his birthright and assume the throne. He founded the town that would bear his name, Rome, and became its first king.

Or so the legend goes. The story was probably made up about 300 BC, some 450 years after Romulus supposedly founded Rome, and there's no chance of it being true. For a start, while it's not unheard of for animals to adopt the young of another species, a wolf rearing two human babies is far less likely than her eating them. And yes, while rulers have in the past committed infanticide, and while the twins' great uncle may well have done so by leaving them by a river bank, or throwing them in the river, even if a she-wolf did pick up the boys and carry them back to her den, her milk would have killed them.

You see, a wolf's milk isn't dissimilar to that of a modern-day dog, and it's way more nutrient dense than human breastmilk. Wolf milk has about two and a half times more kilojoules per gram, 2–3 times more fat (9.5%), 10 times as much protein (8% compared to only 0.8% in humans), and about twice as much sugar.[2] Wolf milk is really good for wolf cubs, but it's way off the mark for what a human baby can handle.

The image of a lactating she-wolf suckling two human babies is the archetypal image of Rome, set in bronze as a statue in the Capitoline Museum. The founders of Siena in Tuscany (the children of Romulus's brother, Remus) were also supposedly suckled by a wolf. In ancient Europe, myths also abounded of people being suckled by bears or other animals. In ancient Persia, Cyrus I was reputedly suckled by a dog, while various later emperors such as Xerxes were supposedly nursed by horses in the centuries pre-Christ. Although Xerxes may well have drunk the milk of mares, it was probably not directly from the horse's teat, as sometimes depicted.

The point is, milk, long recognised as the source of life, has a powerful presence in myth and art, culture and religion.

One of the more bizarre images, courtesy of the Catholic Church, is of the Virgin Mary suckling a man. A very grown man, Saint Bernard, who was born in 1090 in France. A great believer in the importance of Mary's chaste nature, Bernard was apparently in need of reassurance that Mary was, in fact, the mother of God. Pressing his point next to a statue of the Madonna, Bernard called out, *Monstra te esse matre!* – 'Show us to be a mother!' At this point, baby Jesus, who was being suckled, pulled away from his mother's breast and Mary squirted Bernard in the mouth, or the eye, depending on which version of the tale you choose to believe.

As the legend is told, when the breastmilk went into Bernard's mouth, he was gifted with the power of truth-telling – in biblical terms, given wisdom. In the version where the Madonna's milk hit him in the eye, it healed an infection (which, knowing the surprising properties of breastmilk, sounds a bit more probable).

Either way, Bernard can be found in a few depictions from about the year 1200 to the late 1700s – including one by Leonardo da Vinci – pretty much without fail showing the Madonna baring her breast. In most, Bernard is painted monk-like, either fully bearded or fully shaved (including his head), always fully robed, and usually kneeling.

In some renditions he's actually right next to the Madonna, and she has apparently placed him on the breast and is suckling him. In others, a few drops of breastmilk dribble down his forehead. Over time, perhaps authorities started to wonder if their parishioners would see someone creepy rather than pious in the depictions, as Bernard is shown more distant from the Madonna. In one, an oil painting from around 1650 by Alonso Cano, Bernard is a good few metres from the Madonna, and she's squeezing her breast and – somewhat miraculously – squirting her sacred milk straight into his mouth.

Apparently, the real Bernard was a bit sickly, and prone to pleasures of the flesh. He craved abstinence from any temptation. He was anaemic, suffered from migraines, had an inflamed gut, sported high blood pressure, and also had a pretty poor sense of taste.[3] Sadly, history makes no mention of whether the Madonna's milk cured any of that.

Madonna Lactans, as the image of Mary suckling him is called, certainly didn't harm Bernard's reputation. The less generous might argue that he made up the story to aid his career. Which would be odd for a white man, particularly one in the Church, but there you have it. His saintly reputation certainly helped give him credence with all kinds of things, including his ardent support for the Crusades – whose main aim, according to Bernard, was to battle the pagan Slavs 'until such a time as, by God's help, they shall either be converted or deleted'. Bernard's zeal, and that of others who supported the Crusades, managed to cost somewhere between 1 and 9 million lives, so perhaps he could have done with more of the Madonna's milk to truly find wisdom.

Be it lactating Madonnas or suckling wolves, milk is undoubtedly a powerful cultural icon, though its symbolic power certainly varies. While the image of a breastfeeding Madonna could be seen as a celebration of maternity – a mother suckling her young, the natural way of things – an image of a grown man pursing his lips to a lactating woman's breast, a woman not his wife, is perhaps less so.

The fully normal, vital bodily function of lactation has been politicised and commercialised. In the 1980s, the giant multinational food company Nestlé was caught out trying to sell baby formula to poorer countries by body-shaming women. Harry Grebert, marketing manager for an ad agency working for a multinational infant formula company at the time, was interviewed on Australian radio about pushing sales into developing nations. When asked, 'Did you discuss the fact that you were trying to supplant perfectly good mother's milk?', he replied:

> Yes. I have to say yes, that is true. When we were thinking around the ideas of visual imagery, we wanted to use a nursing mother with the youngest possible baby. The strategy was formed around the idea that mothers had better things

to do than to nurse their babies. There was also a cosmetic proposition that was appealing to the idea that if you nursed your baby you might suffer from what was referred to as 'bosom sag'.[4]

Milk's pivotal role as a life-giving force – its vitality, if you will – has long been used in marketing in a range of devious ways, playing on sexual politics, and reinforcing gender and racial divides. From the enslaved wet nurses of the Americas, to the ability to fill factories with working women, and the role it has played in body image, breastmilk has been revered and reviled. From the deific image of mothers reflected in the Madonna, to the strange idea that a woman's milk is an aid to a grown man's wisdom and career, milk is a powerful tool. It taps into our cultural memory, and into our sense of self and how we identify. Far from simply being a perfect superfood, milk has become a cultural weapon, and a racial flashpoint.

Take some followers of Donald Trump, for instance. During the turbulent times around his 2017 presidency, white supremacists were shown chanting anti-Semitic slogans. These activists were chugging bottles of what they considered to be an indicator of their race – milk – in response to a live-stream video by artist Luke Turner, who, on the day Trump was inaugurated as America's president, placed a sign on a wall in New York saying 'He Will Not Divide US', and invited people to respond to the phrase however seemed appropriate to them.

It didn't take long for far-right activists to find the wall-mounted camera Turner had installed, and dance around shirtless, shouting 'down with the vegan agenda' and slurping milk from gallon-sized containers. They selected milk as their manly drink of choice (it was all men, as seems often the case in neo-Nazi circles). Originally planned to film for the entire Trump presidency, Turner's artwork attracted so many right-wing agitators that the Museum of the Moving Image shut it down just 21 days into its four-year exhibition, citing concerns that it had become 'a serious and ongoing public safety hazard'.[5]

Strangely, milk, this most seemingly innocent liquid, is often the alt-right's drink of choice.[6] Richard Spencer, the white power exponent who coined the term 'alt-right', once tweeted, 'I'm very tolerant … lactose tolerant!', while other supremacists have used the term *Heil milk!* as a slogan, indicating – amongst other things – their belief that there is some worldwide Jewish conspiracy to somehow rule or control the globe, a conspiracy that is also a vegan agenda.

Milk as the desirable drink for a strong nation also finds some support from people such as Justin Cook, an assistant professor of economics at the University of California, Merced, who, writing for America's Public Broadcasting Service (PBS) *NewsHour*, suggested:

Could milk consumption have contributed to Europe's colonization of most of the world during the 16th century? The answer 'yes' is more likely than you may think.[7]

The thinking goes that milk-drinking cultures are more likely to be richer, and therefore stronger, than their non-milk-drinking counterparts.

In a 2018 article in the University of Wollongong's *Animal Journal Studies*, titled '"White Power Milk": Milk, Dietary Racism, and the "Alt-Right"', Mercer University's Vasile Stănescu unpicks the use of milk as a symbol of white supremacy, and its continuation of a style of colonial racism that started in the 1800s. Milk's colour, the fact that not all adults around the world can safely consume its lactose, and some questionable genomic reasoning around who colonised whom, argues Stănescu, has led to it being associated with power. And despite parts of Africa and Pakistan having a high lactose tolerance – less rapacious regions that didn't see the need to invade far-distant nations and colonise them – such evidence was ignored by those who wanted to use milk's profile, colour and good name to promote white supremacy.

Milk's role in racial politics and sexual politics is complex and fraught. Sometimes it intertwines, and that's because lactation can also play a role in birth control. For some women, breastfeeding curtails the return of their periods and fertility for over 12 months. American slave owners, keen to raise the next generation of slaves, found this effect problematic, so it wasn't unheard of to deny infants of slaves their mother's breast to speed up her return to fertility, so more slave babies could be born. In fact, there are accounts from the 1500s to the 1800s of men in Europe preventing their wives breastfeeding so they could more rapidly enlarge their number of possible heirs.[8]

Dairy's impact in race politics is sometimes way more subtle. In 2023, the Wellcome Collection museum in the UK ran an exhibition on milk, and a big part of it looked at milk's role in cultural stereotypes. The marketing of milk in English-speaking, predominantly white nations such as the US, UK, Canada, Australia and New Zealand has long favoured the imagery of the nuclear (white) family, glimpses of green rolling hills, and the notion that those who truly love their family feed them milk. Posters such as a 1920 advertisement from the Oregon Dairymen's League and the Portland Milk Distributors suggest: 'The greatest single factor in making Americans a great people, say scientists, is the habitual use of milk and milk products.'

Propaganda in the early 1900s was rife. The National Milk Publicity Council in the UK used photographs to demonstrate increased growth in children who drank milk, with those supposedly bigger and stronger as the milk drinkers. The images were dodgy, however, because they used only boys from low-socioeconomic backgrounds, and misrepresented milk's role in growth. The Medical Research Council of the time asked for them to be withdrawn.

Milk was constantly on the fringes of questions of race. In the US in the 1920s, Herbert Hoover, who later became US president during the Depression, was quoted in butter advertisements as saying, 'The white race cannot survive without dairy products.' It's a quote from his essay in the 1922 edition of the trade press, *Dairy World*, while Hoover was in the powerful role of US Secretary of Commerce. Hoover was a strong believer in eugenics – the theory that some races are simply superior to others. He backed his views in print with a speech at the World Dairy Congress in 1923:

> **Upon this industry, more than any other of the food industries, depends not alone the problem of public health but there depend on it the very growth and virility of the white race.[9]**

Milk isn't just a drink, a source of nutrients. It's a cultural icon or gastronomic flashpoint. It has resonance beyond the mere physical. But it's the physical that should matter more in debates about milk.

As we know, not everybody can drink milk with impunity, and not everybody needs to – or should – eat dairy. But before we get to when, how and why we might want to use animal milk as adults, to understand exactly what milk is, and why it matters, we first need to look at the milk we're designed to get in early life – and just how miraculous, and fraught, even that process is.

Motherly Instinct

There was a frantic, urgent mooing from the paddock. Multiple cows were bellowing incessantly. Something was going badly wrong in one of our south-facing fields.

I raced to see what was happening. When I arrived, I found a freshly born calf lying in the grass, half the mob circling it, much of its birth shroud still attached. It was alive, but yet to stand. The calf's mother, Jasmine, who had birthed four times successfully prior to then, was the loudest of the moo-ers. She snorted, angrily, like a bull. She lowered her head, and as I ran up to see what the commotion was all about, Jasmine charged at her calf, crashed into her daughter and knocked the little one down the hill. Jasmine thrashed her head about like a headbanger as a warning, then ran down the hill away from me, and promptly tried to stomp on her daughter.

That is nature in action. Jasmine, an animal practised in mothering, was actively trying to kill her newborn. Olia, as the calf was soon named (after the fantastic Ukrainian cookbook author, Olia Hercules), was puzzled. No sooner had she drawn breath than her mother was trying to murder her.

We like to think of the mothering instinct as the normal way of things, that the bond between mother and child is something sacred, sacrosanct and guaranteed. But mothering doesn't always work like that, with humans just as with cows. Sometimes things just go awry.

For three months the only way we could get little Olia a drink from her mum was to tie Jasmine up and force the calf on, avoiding the kicks thrown at us and the calf. Then, after a few weeks, we had to supervise Jasmine in the paddock while Olia would feed. Despite surviving the ordeal, Olia remains stunted as a result of not getting milk on demand.

The marvel of milk feeding relies on a few things. It relies on the mother wanting to suckle her young, obviously. It relies on the mother's body being able to produce enough milk. It also relies on the ability of the baby to drink, and on conditions being comfortable enough for both mother and young to nurse successfully. Simple things such as stress, hunger, heat and cold can affect a mother's milk supply and letdown – as can any number of biological causes, from serious illness to reasons unknown.

On our farm, we've seen a few problems with what should come naturally. There have been a handful of cows that won't suckle their young, like Jasmine. Some cows have stopped suckling their young, like our dairy cow Alice, whose calf died a fortnight after being born, despite supplementary feeding when we noticed it flailing. We've had calves and goats and sheep that just don't really know how to get to a teat, or are too frail to, such as Ailsa, the kid goat we met earlier. In nature, these young would be picked off by carnivores. In nature, mothers abandon their young when threatened by a predator. Pademelons, a type of wallaby we see a lot of on our farm, have been known to toss a baby out of the pouch when stressed or chased, a survival strategy that seems barbaric to us. Rats, a species that suckle their young and sing to their babies, can eat over half the pups they produce if they're stressed. I've seen pigs cannibalise their babies when disturbed. Historically, we humans have done terrible things to our children, both willingly and as a result of impaired biology.[1]

What we modern humans think of as mothering is neither as entirely natural, or as easy, as we like to think. We've always had difficulty feeding babies.

Milk is made by a very complex organ, the mammary gland. And a warning to the squeamish: in this section we'll be talking a lot of milk biology – where it comes from and how it comes out.

In the blue whale, each mammary gland is the size of a baby elephant.[2] Between her two glands, that adds up to a quarter of a tonne of whale.[3]

Some animals, such as pigs, have more than two mammary glands, with about a dozen nipples – even up to 16, which is handy when modern pig breeds can have up to 14 babies at a time.

But even they look like they're not trying compared to the tailless tenrec, a Madagascan insect-eater that clocks in with 36 teats to suckle up to 30 babies.[4] A different survival strategy than the Tasmanian devil, that's for sure, whose 30 babies, as you may recall, all compete for just four teats.

The miracle of how milk is made, and therefore all the things that can go wrong, is quite complex. Essentially, a mother's body will make up a milk designed for her young – not just tailored to her species, but to the age of her young, and in some cases, such as some primates, to her baby's gender.[5] What's more, the components of this milk might vary a little or a lot over the course of lactation.

Louise is being very forthright. She's talking nipples, and areolas, and mastitis. She's frank about boobs, uteruses, hormones, breasts. She swings from medical speak to slang, from idealistic to pragmatic.

I'm talking to Louise Klug about the biology of breasts because she milks humans for a living. Well, almost. She's a midwife and lactation consultant, and while I'm in no position to lactate, I'm keen to know what difficulties a mother encounters, and just how hard it is to do something that supposedly comes naturally. Louise and I were introduced by friends after the premature birth of their daughter.

Louise works in the hospital system and the community, with mothers from all walks of life. About 70% of first-time mums see lactation consultants such as Louise, who help them to get the baby on the breast. After that initial week, something like 30% of mothers still have difficulty (or think they have difficulty) making milk, or getting the baby to take it. In the animal kingdom, if nearly a third of mothers were having trouble nourishing their young, it'd be a genetic dead end. Are we humans different?

Talk to pretty much any first-time mother, and they'll tell you that breastfeeding isn't something that just happens easily. From how to sit, to how to latch the baby on, and how to tell if the little one is just playful or hungry, the process is fraught. Then there are issues with milk production, milk letdown, cracked nipples, infections, feelings of worthlessness, the pressure of being a proper mum.

Some of this difficulty is a result of how good we have become at helping to keep mothers and babies alive, and with advances in fertility. As conception and birth become more invasive, with things such as induced births and donor eggs and sperm, Louise thinks some mothers have lost confidence in their bodies to do the right thing. Lost their intuition. It's her mission to try to correct some of that.

Much of what Louise does is point out some misunderstandings about suckling a baby. Grabbing a crying baby from their cot, undoing the maternity bra and popping the baby's mouth on the breast might work, she says, but that's not the way nature designed it. She suggests a mother should pop a hungry baby between the breasts, so her boobs touch the infant's cheeks, where the newborn is hardwired to turn towards the source of their food. Babies tend to do it with an open mouth and with a seeking nature. They make an up-and-down movement with their heads, that looks like snuffling, which helps the mother let down milk and prepares the baby for the feed. They then put their tongue down and forward, ready to find the nipple.

A baby left to its own devices on a mother's chest will usually find a drink, Louise says. Milk is so important that babies are hard-wired to the smell, not only of the milk itself, but also the breast. Things called Montgomery's glands line the areola, the coloured part around the nipple, and they exude a few things, including a mother's individual scent. When babies get a whiff of this, or of milk and colostrum, it can elicit a crawl response in a newborn, they're so keen to get hold of the good stuff.[6]

This behaviour is mirrored in all mammals. Piglets find their mother's teats through smell, too, and by following her hair patterns.[7] Calves find the teat through trial and error, using their mother's smell, size, shape and vocal cues, along with the mother's stance as she licks her calf to help them locate the udder.[8] Foals use visual cues and scent, along with gentle nudges from their mother to find the teat.[9] Echidnas don't even have nipples or teats, so the babies find the exuded milk dribbling out from glands just under the skin. Dolphins don't have external nipples, but rather slits that hide them, that the baby has to nudge. And bats have a spare set of nipples that aren't there to drink from, but rather to hang onto when the mother bat flies. All these animals have to drink milk, and to drink milk early and often, to have the best chance of survival.

When the process goes well, all is well. But when the process goes wrong, it can go horribly wrong. A mother may not produce milk, or produce too much milk and risk infection. A human baby may be tongue-tied, where the tongue is held from poking out by a strip of skin at its base, meaning the bub can't take the nipple properly into its mouth. The baby may have a cleft palate. The mother may get infections. Rejection by the mother, an inability to suck by the baby, or a low supply of maternal milk are all real problems for all kinds of animals, humans included. Breastfeeding for some can be very hard. It takes time, effort, consciousness and presence. For some, as the lauded 19th century British writer Isabella Beeton wrote in her seminal tome, *Mrs. Beeton's Book of Household Management*, the period of suckling is 'the season of penance to the mother'.[10]

The way nature *should* work is that milk formation, the hard-wiring of the mother, starts in early pregnancy. She can even produce colostrum from as little as 23 weeks into the 40 weeks' gestation. As the mother is due to give birth, certain hormones are released from the pituitary gland, hormones that start the production of milk (for more on this, see the chapter How Real Milk is Made). The main hormones in humans are prolactin and oxytocin, which together enlarge the breasts, and fill up the mammary glands with colostrum, and later with the actual milk. This allows the mother to produce enough milk for her baby, or babies.

The early milk is produced automatically, through the endocrine system, a hormonal system embedded in the mother. In humans, as with cows or sheep or goats and most other mammals, the mother produces enough milk for the potential number of babies. For us, that is usually enough milk for two babies, to allow for the birth of twins. For a pig, that's enough milk for 14 babies. For a kangaroo, it's just enough for one, as they already have an older joey feeding on their only other teat.

Later, a few days in, the milk production is controlled by the autocrine system, a localised way to regulate how much milk is made, and what it contains.

When a baby starts to suck, the rhythmic action of feeding triggers milk production. Essentially, it's a self-sustaining system; the more the baby drinks, the more milk is produced. Hence, while women usually start out producing enough colostrum and then milk for two babies, and hence can feel uncomfortably full, milk production drops to suit the most common outcome, a single baby per pregnancy – because that's all the milk that is being drunk.

For humans, the milk is delivered via the breast, but it's not like a cow or sheep, where a single hole can squirt out the milk. On average, the nipple of a woman has nine holes, milk ducts, that all release milk to the outside. Some women have 20 milk ducts per nipple. The areola, that textured part around the nipple, is important too, in part because of those Montgomery glands we met earlier. They secrete oily substances that help protect the nipple and make suckling the baby easier.[11] These oils are also the ones that contain the mother's scent, which attracts the baby to the breast. This smell lowers the arousal state in active newborns and increases them in sleepy ones, while at the same time it also helps the baby do more positive head turning and stimulates their appetite.[12] In other words, the scent helps excited babies relax, ready to feed; it helps drowsy babies wake up for feeds – and it helps babies to turn their head to look for food. Pretty cool, hey?

Proper feeding means that most (about two-thirds) of what a baby latches onto is the areola and breast, and only a third is the nipple. And this is important, too, because stimulation of the areola causes the mum's body to produce more prolactin, which peaks about half an hour after feeding. The amount of prolactin a woman produces increases at night, and is the trigger for the body to make more milk. It also helps suppress ovulation, a useful form of birth control (on a population level, but very unreliable on a personal level) – though this effect is also most pronounced with night feeds, for some reason we don't yet fully understand.

Oxytocin is the hormone of milk letdown. A woman needs to be relaxed for the milk to flow, and stimulation of the nipple triggers oxytocin release. We know that oxytocin is released earlier than prolactin in women, thanks to things such as when the mother thinks lovingly about her child, hears her child cry, cuddles her child and more. That's why the milk letdown in mothers can be triggered by the sound or sight of someone else's baby. And why one breast can leak while the other is being suckled. But this letdown can be hampered by stress; physical or emotional. In a nice feedback loop, however, the release of oxytocin reduces stress in the mother and increases the feelings of love between parent and child. In essence, it increases the mothering.[13]

Milk letdown seems to be universal. Even commercial dairies rely on the letdown reflex, getting cows into routines, making them comfortable and feeling safe to let the milk flow as the machine pumps. We also know that certain music is helpful for oxytocin release and hence milk letdown, including in cows. Playing them R.E.M. or classical music is better than hard rock or rap, apparently, and not only helps with letdown (oxytocin) but also leads to higher milk production (prolactin).[14]

Letdown is a real thing, as any mother can tell you. It's a complex biological mechanism that involves both the milker and the milked. The thing is, a cow – like any mother – has the capacity to hold back her milk. One of our dairy cows, Ruby, a good mum by any standard, will allow us to milk her each morning after 12 hours away from her calf. She can give us 10 litres (2 gallons) of milk, seemingly without bother, before she even looks for her calf. But what she gifts so easily is the early milk, the foremilk as it's called, not the creamy rich part. So her milk sometimes measures less than 2% butterfat until we wean the calf after six months or so. Then, miraculously, with no baby to save her milk for, she produces more milk for us – and, importantly, the fat content rises to 4%.

What we see in our micro-dairy is true of most mammals. The milk an animal delivers varies by the time between each feed. The first milk the baby drinks, the foremilk, can be thinner than the later milk, known as hindmilk. One study has shown that the hindmilk a human baby receives could have 2–3 times the fat content of the foremilk.[15] It's not that the milk glands produce different milk over time, or that there are different compartments for milks of different nutrient density. It's just that, in the same way cream rises to the top in unhomogenised milk, the fat separates out with time; the creamier milk sticks to the walls of the mammary gland more, so it comes out later. That's one reason it's recommended that human mothers empty one breast before starting the baby on the other, because her body intuitively knows what to make for her child, but the baby can only get that magic cocktail of nutrients if it can drink all the milk provided at one time.

For most of history, if a human mother had trouble feeding a baby, there was a fallback position. But as time went by, that became more fraught – so much so that, as we'll see in the next chapter, it once caused a whole city to become a baby-free zone.

The City
Without Babies

Audrey is old, for a sow. She's about twice the age a commercial pig would ever get to. Her age means she's less fertile. However, one newly born slip, as baby pigs are called where we live, was born only half formed. This can happen in large litters where the pigs are lined up *in utero*, the last often smaller and getting fewer nutrients from the placenta, and it's more common in older mums. Having eight live slips, however, is about average for our farm, so we didn't stress unnecessarily. Audrey was fat and healthy looking. She'd had several litters before, and she's a really great mum. I watched as she lay on her side, very, very carefully so she didn't crush her babies, nudging them away as she flopped down, exposing two rows of teats for them to feed from.

The piglets rooted around, each finding a teat, which would be their personal nipple for every feed, and they started to bobble with their noses to let the milk down. Eventually, each one latched on, allowing Audrey to suckle her brood; each piglet getting a table-spoon or two of milk each feed, every 45 minutes, day and night.

A week later, however, and I was worried. Those eight piglets looked skinny. The runt of the litter – and there's always a smaller, weaker slip called a runt – was fading. It wasn't thriving. The runt always ends up at the back end of the sow, less able to fight for the front teats. At the rear, the milk supply is lower, and only has half the fat, too, so the runt falls further and further behind.[1] Piglets of all sizes usually survive if they can get through the first 48 hours. This one, however, despite a last-ditch attempt to keep it alive with eye droppers of cow colostrum, died.

There was an issue with Audrey, too, we thought, as she was no longer thriving. The vet diagnosed postpartum dysgalactia. At the time, I thought it an odd name for a disease, sounding like it was named for the stars. I now know, of course, that *galactic* gave its name to our galaxies, and dysgalactia means 'disrupted milk'. Audrey could feed, but not feed her litter properly, or as abundantly, as her little ones needed.

Pain relief, antibiotics and oxytocin for milk letdown all cured Audrey, and allowed her to produce plenty of rich piggy milk for her remaining litter – though giving three injections twice a day to a 200 kg (440 lb) sow with decent tusks is something I don't recommend. On bigger pig farms that have sheds, where dysgalactia is more common, they can simply shuttle babies around between mothers after birth, putting the runts all on one mum, the bigger ones on another, to even out the growth, and to cope with mothers who have trouble feeding their young.

Audrey isn't the only mother to fall ill and have trouble feeding her babies. Plenty of animals go through it, humans included. We also have the option of not feeding our own babies, too, which – prior to the invention of infant formula – led to the rise and rise of the wet nurse.

A wet nurse is a woman who breastfeeds someone else's baby. Arguably the idea has been around since the dawn of time, because a community raises a baby, and if a woman was ill, if her milk dried up, or if she died in childbirth, a baby could be raised by another lactating woman.

Wet-nursing can also be done to species not our own. In 2014 an American woman posted pictures of herself suckling five orphaned puppies to keep them alive and received all kinds of vitriol in response.[2] But her actions are nothing new. Many cultures would use a nursing mother to keep other animals alive, from Native Americans to Australia's Aboriginal people. In his book *My Crowded Solitude* from 1926, Jack McLaren recounts often seeing Indigenous women suckling orphaned dingo puppies; so vital was the dog for hunting and warmth that surplus breastmilk was used to keep them alive.[3] In Kamchatka, bears were suckled by the Itelmens people because their meat was so prized. In the US, some women used hungry puppies to harden their nipples.[4] Mary Wollstonecraft, a British writer who was dying of a fever after the birth of her second daughter in 1797, was encouraged to suckle puppies to help her expel her infected placenta, a strategy that sadly failed.

The main animals that human mothers fed, however, has always been humans, even if they weren't their own children. Before we started congregating in large communities after the agricultural revolution, about 10,000 years ago, wet-nursing human babies was clearly a survival strategy for the clan. But as population centres grew, the use of the wet nurse led to huge social changes, so much so that at one stage in the 18th century, Paris became known as the City Without Babies, as nine out of ten Parisian babies were sent out to wet nurses in the country.

As we saw in the last chapter, feeding babies isn't always easy. Biology fails. People get stressed or ill. And for a long while, the whole birthing process was dangerously fraught, with many women dying in the hours and days after giving birth. Add to that mastitis, other illnesses, mental despair, poverty and the like, and many mothers simply couldn't care for their babies. The idea of the wet nurse, then, is noble. In reality, it became far from it as human milk – in human form as the wet nurse – became commodified.

Wet-nursing, naturally, could be a foundational beginning for a child – but it could also bestow highly beneficial social rewards. In China, the women who wet-nursed emperors were given honorary status, and became known as the Nurse Empress Dowager. Most Mughal princes, royalty of the subcontinent from the 1500s to 1800s, were wet-nursed too, conferring kinship between the nurse and the nursed through Muslim law.[5] A 10th century Quranic scholar, Al-Tabari, cited the prophet Muhammad on the topic:

> **She receives, for every mouthful and for every suck, the reward of one good deed. And if she is kept awake by her child at night, she receives the reward of one who frees seventy slaves for the sake of Allah.**

In Japan, the famed Kasuga no Tsubone (Lady Kasuga) wet-nursed a shogun nobleman, and became a powerful political operative in the process.[6]

But while wet-nursing can undoubtedly be altruistic and noble, as always, human nature, the free market, and the power differentials between the rich and poor are ever-present.

What happened in reality with wet-nursing over the years is that impoverished mothers neglected their own children to feed richer people's babies. Or, after expressing their milk, they watered it down to make it stretch further, depriving both babies of nourishment. The relationship needs to be built on mutual trust and respect, but often was based on a power and wealth imbalance.

In *Mrs. Beeton's Book of Household Management*, which came out in 1861, Isabella Beeton suggests there is nothing in domestic management 'more fraught with vexation and disquietude' than employing a wet nurse.[7]

She goes on to describe desirable traits in the wet nurse. They should be 'sound in every respect, and the body free from all eruptive disease or local blemish'. She should be ruddy in skin, have full, round and elastic breasts, without 'pendulous, flabby and relaxed' nipples. If the nipples are pendulous, the 'milk is sure to be imperfect in its organisation, and, consequently, deficient in its nutrient qualities'.

Mrs Beeton goes on to give quite specific instructions about diet:

> Respecting the diet of the wet-nurse ... The food itself should
> be light, easy of digestion, and simple. Boiled or roast meat,
> with bread and potatoes, with occasionally a piece of sago, rice
> or tapioca pudding ... broths, green vegetables, and all acid or
> salt foods must be avoided ... If the dinner is taken early –
> at one o'clock – there will be no occasion for luncheon ...
> Half a pint of stout, with a biscuit, at eleven o'clock, will be
> abundantly sufficient between breakfast at eight, and a good
> dinner, with a pint of porter, at one o'clock. About eight o'clock
> in the evening, half a pint of stout, with another biscuit, may be
> taken; and for supper, at ten or half-past, a pint of porter, with
> a slice of toast or a small amount of bread and cheese.[8]

It seems that boring food, a pint of stout and two pints of porter were just the ticket to wet-nursing perfection.

While wet-nursing was common in Britain, nowhere was it as common as France in the 1800s, where a wet nurse was called a *nourrice* – a word deriving from the Latin *nutricius*, meaning 'nourishment', but also 'to suckle'. *Nutricius* also gave us 'nutrition' and 'nurse', because to nurse used to exclusively mean to suckle a child.

Before wet nurses really took off in Paris, it was generally only the aristocrats who could afford to actually hire one. Louise Bourgeois, midwife of Queen Marie de Medici, was famed for being the first European woman to write of her craft, and in 1609 also wrote a pretty interesting treatise on childbirth, entitled *Diverse observations on sterility, miscarriage, fertility, childbirth, and diseases of women and new-born children*. In it, she makes Mrs Beeton look like a pushover on what to look for in a wet nurse:

> The important thing to consider is her gaze, such as whether
> she looks directly at you, is cross-eyed, or looks downcast ...

Take care that she is not a redhead, because their milk is very hot … Observe whether her teeth are white and well set … Find out if any bad odour comes from her nose, for the least strong smell emanating from a wet nurse's nose or mouth greatly harms the child's lungs, in the same way that the vapor rising from mud or a privy can spoil bronze, copper, or silver and blacken it … A wet nurse should therefore be pleasant, have good teeth, dark or brown hair, and come from a healthy family … She should not be choleric; she should have good, abundant milk. Her nipples should not be too thick … She should not be too fat, and above all, make sure she is not of an amorous disposition. This is often the case with honest women whose disposition causes them to lie with their husbands. Their milk is then true poison for a nursing child … Truly good wet nurses never have [their periods] while nursing, or at most they have them fifteen or eighteen months after giving birth. I have observed that when they have them earlier, the children languish from that time on.[9]

Such prescriptive notions of the ideal wet nurse must have made finding the appropriate woman to feed your child almost impossible – especially as the demand for wet nurses grew. According to Jean-Charles-Pierre Lenoir, a lawyer who headed the Parisian police just prior to the French Revolution, of the 21,000 babies who were born in Paris in 1780, barely 1000 were nursed by their own mothers.[10] And three-quarters of those babies were sent to the countryside to be nursed by other women. For aesthetic reasons, and the fact it was considered beneath women of a certain class, it simply wasn't the fashion to breastfeed.

Irish doctor Hans Sloane (who some erroneously credit with having invented chocolate milk) showed in the mid-1700s that dry-nursed babies (those without a lactating woman to feed from) were nearly three times more likely to die than wet-nursed babies – 54% mortality compared to 19% for wet-nursed babies.[11] In the 1740s, infant mortality sat at about 35% – about one in three babies dying – from all causes,[12] so while sending children out to wet nurses improved their chances of survival over no breastmilk at all, it was certainly no guarantee.

Worry about wet nurses passing on less desirable character traits was a real fear in times gone by, though fashion overrode those fears in Paris and beyond. Even today, despite it being a biological necessity, programmed into our DNA and the way things have happened since time immemorial, breastfeeding still somehow courts controversy and makes some people uncomfortable, or is an anathema. As I was writing this story, an Australian judge asked a breastfeeding mother to vacate his court, not because the baby was being unusually difficult or loud, but because he simply didn't think it appropriate, saying his comments should be 'self-explanatory':

Madam, you will not be permitted to breastfeed a baby in court. I'm sorry. I will have to ask you to leave. It will be a distraction for the jury at the very least.[13]

Many women's working lives often don't allow for the very normal, very common, very necessary act of breastfeeding a baby – whether in public or not – even though it is such a fundamental right. This is a problem mostly for the working poor, as they usually spend longer commuting to work, have less flexible working hours, and may not get maternity leave or have enough savings to allow them time off work. And they may hold more than one job at a time.

The same could be said of Paris in the 1800s. French wet nurses were all from the lower classes – but of course, not all *nourrice* were given work by the rich. Some earned a living as wet-nurses for the *nourrice* in Paris, who, in order to work for rich mothers, also sent their babies out to the countryside to be nourished by others – another reason Paris became known as the City Without Babies.

It's hard sometimes, looking back, to realise how much less we valued human lives in times gone by – and how hard human lives really were. Out of 66,259 Parisian babies sent to country wet nurses between 1770 and 1776, a third were dead by six months. Figures from over a century later, in 1889, show that only a third of Parisian babies placed with wet nurses were actually being breastfed, which dropped to a dismal 7.5% in 1913 thanks to bottle feeding.[14] Poor mothers simply didn't have access to the necessary nourishment to produce the breastmilk needed.

In the mid-1800s, the infant mortality rate was 30–60 times higher than it is today in developed nations.[15] Yes, mothers and fathers grieved for their lost children, but death was common, especially in early life – and death from a lack of the right milk was a major cause, particularly in the first year after birth.

If we go back even further, the statistics paint an even grimmer picture. Between 300 BC and 400 AD, when the Roman Empire was at its height, contracts were drawn up with wet nurses to feed abandoned infants. An article in *The Journal of Perinatal Education*, titled 'A History of Infant Feeding', observed:

> **The infants were usually unwanted females thrown onto rubbish piles. The wealthy purchased the infant as an inexpensive slave for future use, and the wet nurses – who were slaves themselves – fed the infant for up to 3 years.[16]**

The Romans weren't the only people to use slaves as wet nurses, you'll not be surprised to hear. In the US, white slave owners were quite happy to farm out their children to their black female slaves. Or more accurately, quite happy to bring black slaves into the house to feed the slave owner's children, while also preventing the black mothers feeding their own children in the process.[17] This was, apparently, not uncommon: according to a 2022 story in *The Washington Post*, in 1850 about 20% of white enslavers used slave wet nurses – amounting to roughly 70,000 women.

Many modern commentators note that the black 'mamma' used in popular culture and in advertising (often to promote products to white people) comes from this idea of the overtly maternal black woman. The irony is that for most, the choice of feeding a white child was not theirs to make, and was usually at the expense of their own biological children.

Wherever there are slaves, there is often enforced wet-nursing. In 19th century Brazil, for instance, advertisements such as the ones below were commonplace, posted by slave owners renting out young women as wet nurses:

> **For rent, an eighteen-year-old girl, wet nurse, healthy, and with much good milk for the last two months. She is for rent because her child has died. Posted in the Rua da Candelaria No. 18A.**
> *0 Mercantil*, **Rio de Janeiro, April 30, 1845**[18]

Or this:

> **For rent, a wet nurse with very good milk, from her first pregnancy, gave birth six days ago, in the Rua dos Pescadores, No. 64. Be it advised that she does not have a child.**
> *Jornal do Comercio*, **Rio de Janeiro, December 10, 1827**[19]

Be advised, she does not have a child. Let that sit with you for a moment.

With the advent of modern refrigeration, wet-nursing has evolved. In 2015 Ambrosia Labs, a company from Utah, set up breastmilk-collecting clinics in Cambodia, hoping to export breastmilk back to America for people to buy. The UN, however, wasn't impressed and Cambodia subsequently banned its export.[20] Not long after, in 2017, a company called Neolacta was given permission to sell breastmilk garnered from Indian mothers to people in Australia, though the Indian government has since attempted to shut down the trade.[21]

Imported breastmilk aside, the wet nurse is a thing of the past in most of the world today. But there are times when people do breastfeed another's baby. One is with 'adoptive lactation', where a woman who hasn't given birth trains her body to lactate. From all accounts it takes quite a bit of dedication, and some people use hormones to trigger milk production, but some can do it by breast stimulation alone. Regular breast pumping can trigger the same response as a suckling baby, and if done for long enough, and often enough, some women can breastfeed an adopted baby or one born through a surrogate, without the need for formula. If spontaneous breastfeeding after giving birth sounds hard, this sounds really hard, and most people need the help of a specialist lactation consultant.

There's another way people can now access breastmilk. All over the world there are breastmilk banks, where women who have too much milk donate it for those who can't produce enough. There's a handful of such banks in Australia, where the donated milk is reserved for the most vulnerable – the babies born prematurely and at most risk. In Brazil, the undoubted leader in this space, 228 breastmilk banks have provided supplementary milk to mothers and helped reduce infant mortality by 73% since 1990. Europe has nearly 300 such banks, and there are about 750 around the world.

Outside of the formal system, of course, lies the informal system. While breastmilk banks check the mother's health status and pasteurise the milk, there's an underground community swapping or trading or selling milk. Facebook groups exist where people post a need, or a surplus, and the pick-up and use of milk is usually done on trust. Usually no money changes wallets, and mostly the milk ends up in the hands (and bottles) of needy mothers and the mouths of needy babes.

But sometimes it doesn't. Some ends up with those who have a breastmilk fetish, a need like Saint Bernard to be nourished by something they should've outgrown. Apparently, it's a thing.

Others, body-builders included, see breastmilk as something vital, perhaps supernatural, that is designed specifically for humans and has nourishment beyond anything else you can buy. There are also those who believe breastmilk's life-affirming, immune-boosting responses could be their last resort to cure their maladies, terminal cancers included. Some reports suggest people are paying $500 a litre for breastmilk on the black market. That market includes athletes looking for a competitive edge, all the way to people with more nocturnal athleticism in mind – those with erectile dysfunction.

Breastmilk ends up in all kinds of places. In the UK, there's been breastmilk ice cream for sale, and in the US, a sweets company makes a lolly that – apparently – tastes like breastmilk.

Unregulated breastmilk does pose risks. In one analysis, 74% of human milks bought online contained Gram-negative bacteria, the bad bacteria that are more likely to be antibiotic resistant.[22] That same analysis also had a sobering list of nasties that could be present in illicitly bought breastmilk, including:

> a host of infectious diseases, including cytomegalovirus,
> hepatitis B and C, HIV, HTLV-I&II [viruses that can
> potentially cause leukaemia] and syphilis.

Reading that, I'd want to know where the ingredients for my breastmilk ice cream came from.

Massive advances have helped keep way more babies alive over the last 150 years, and those advances include leaps in hygiene and medicine. And another great advancement has been a real game-changer in terms of how we feed our infants – not just *what* they can be fed and how good it is for them, but also *who* can feed them.

Few things have altered the way we feed babies as much as the advent of baby formula.

Formulaic

I'm always wary of giving our good friend Suzy Manigian a hug. That's because ever since I met Suzy, a dozen or so years ago, she's pretty much always been hiding something living down her front.

What Suzy is usually snuggling against the warmth of her chest is an orphaned marsupial. One day, it was Princess Fiona.

Over the years, Suzy has mothered the young of many of Tasmania's wild creatures. Since 1990 she's been a wildlife carer who takes on the babies of animals killed on roads, on farms, or in other ways. On average, 57 animals die every hour on Tasmania's roads, and many of them have joeys[1] – so the services of people like Suzy are never short in demand. Most rescued joeys are still of the age that, in nature, they'd be pouch-bound and have to be milk fed.

Suzy found Princess Fiona on the Midlands Highway in January 2022, when she checked a dead wombat that had been hit by a car. Inside the moist, impossibly soft, rear-facing pouch was a nude-pink baby latched onto the teat. Fiona, already a couple of months old at that stage, weighed only 270 grams (9 oz). That's slightly more than a pat of butter where I live. Like a big avocado, Suzy reckons, considering a wombat's shape.

In the wild, Princess Fiona would stay inside her mother's pouch being nourished solely by milk until she was six months old. She'd then keep drinking off mum for over a year. At eight weeks old, she is barely dusted with fur. The bones in her skull are still forming, so her ears stick out horizontally from her head; she was named Princess Fiona after the character in *Shrek*, who shares her side-ear feature.

In those first few months, while still blind and deaf and with plenty of her body yet to fully form, Fiona would die without milk. But wombat milk is markedly different from ours, so human formula won't do. Neither will the milk of cows, or sheep, or goats – mostly because they're all too high in the milk sugar lactose, but also because they're not the perfect food for a wombat in other ways. In the 1960s and 1970s, people used 'Carnation' milk as it was branded – evaporated milk (sometimes called unsweetened condensed milk) – to keep baby wombats alive, with limited success. These days, it's so much more scientific. Suzy uses a product from Wombaroo, a specialist marsupial formula maker.

Wombaroo's wombat milk is made using low-lactose cow's milk, with added proteins, oils, vitamins and minerals. Whereas cow's milk is about 3.5% fat and 3% protein, wombat milk is about 10% fat and 8% protein. But while cow's milk has about 5% sugars, wombat milk is virtually sugar free.

In place of a pouch, which is humid and warm, and helps the animal to socialise to another being, Suzy uses flannelette bags that she tucks into a sling that sits under her jumper on her chest. Fiona should drink no more than 10% of her body weight per day, spaced over eight or more feeds. That means she can have about 3 millilitres of formula at a time; just over half a teaspoon at every feed. She's suckled every 2–3 hours, including overnight. Where once she could drink at will, now she must wait until Suzy feeds her.

While wombat formula may have much of the nutrition Fiona needs, it's no mother's milk. Wombat milk, like all milk, confers immunity on the newborn, and so a week or two after bottle feeding starts, Fiona's condition, like most marsupials that go on formula, drops off a cliff. Formula feeds her body, but it doesn't feed her immune system. Luckily, Fiona's immune system was already advanced enough to allow her to bounce back.

When I talk to Suzy 18 months later, Fiona is a whopping 12 kg (26 lb) and nowhere to be seen. She's in wombat daycare when Suzy goes to work.

Wombaroo does amazing things to keep Australian native animals alive pre-weaning. They've analysed the milk of bandicoots and echidnas and bats to come up with a suitable milk replacer for each. They have made formula for pandas, and even a baby hippo. All of these animals would be better off with their mother's milk, but these substitute milks have kept orphaned or milk-deprived young mammals alive. All of these incredible advances have come about because of the leaps we've made in understanding nutrition.

And because marsupials are born so unformed, our attention has turned to considering just what is in those marsupial milks, wondering if they might hold the clues to helping premature human babies thrive.

Before formula, women who had just given birth had little choice when it came to nourishing their newborn babies: either breastfeed, find a wet nurse (not an option for the vast majority), or use another animal's milk. For centuries, people fed their infants with things they thought would nourish them, with no or poor effect. Because breast-milk tastes sweet, they would make a mixture that included honey, or sugar – but because a human baby can't digest those sugars, that just made the baby sicker, hence the high death rate recorded from feeding babies sweetened condensed milk. Others made mixtures using flour, because mixed with cow, sheep or goat's milk, such mixes are known to be relatively nutritious for adults. Over time, babies were fed concoctions based on animal milk and grains – perhaps bread mixed with milk, called *pan* (from the Latin, meaning 'bread', as are the French *pain* and Italian *pane*). Or the mixture might use grains such as oats or wheat, perhaps ground, then boiled with milk, called *pap*. But none of these really nourished a child. They might help keep a baby alive, but they didn't help it thrive.

Dr William Howarth, writing in the medical journal *The Lancet* in 1905, found that the death rate in breastfed babies should be about 65 per 1000 live births in the UK – but for 'hand fed' infants (those not on any breast), that rate was over 200 per 1000 births.[2]

Essentially, one in five babies died if not breastfed.

Most cultures where grazing animals were milked settled on that as a substitute, with or without starches added. In cultures where animals weren't milked, other solutions were developed.

Records from 1725 give an insight into how other societies coped. A mother of six, Elizabeth Hanson was a colonist in New Hampshire who was abducted by Native Americans during the Wabanaki– New England War. Two of her older children died in the capture, and a fortnight later, while still captive, her milk started to dry up, probably from hunger and stress.

Distraught, and with her one-month-old baby daughter wasting away, Elizabeth fed the girl with what she had to hand, including broth from beavers and beaver's guts. At one point, she dribbled water onto her breasts just so the baby had something to drink. Elder women from the group who had abducted her, the Abenaki, saw this and showed her their way to make formula. It involved pounding walnuts into a paste, then mixing the paste with ground corn, which was then boiled. This watery mash revitalised her baby after Elizabeth's milk dried up. Apparently, this technique of making infant food had long been a tradition of the Wabanaki Confederacy, as the group of local Native American nations is known.

The oldest surviving implement used to feed babies was made about 8000 years ago,[3] crafted from clay and shaped a bit like a jug with a spout. The earliest baby bottle that we know for sure contained ruminant milk is 3000 years old. We've been feeding babies milk from other animals for thousands of years, but in more modern times, because of the change in the way we treated animals, and milk, this practice became way more dangerous.

The agricultural revolution, starting about 10,000 years ago, had set the scene for us to milk our grazing animals. But once we brought larger and larger groups of humans together, thanks to the Industrial Revolution starting about 250 years ago, that once vital ruminant milk wasn't as fresh, or as clean, as it had been, and there was a rise in children dying from malnutrition. As we became more urban, the risks blew out.

Increasingly, milk came from further away, or came from animals kept in appallingly crowded or dirty barns. In the US, the early to mid 1800s saw the rise of urban breweries and distilleries, which generated a lot of waste in the form of malted barley. For every litre of beer, you end up with 200 g (7 oz) of spent malted barley, which was fed to animals – and in this case, it was fed to cattle. Cows that were kept and milked in barns. Cows that never saw or ate fresh grass.

Eventually called 'swill milk', this dairy product was produced in sheds where there was no fresh air, no exercise for the cows, and tuberculosis infection among the animals was rampant. Up to 80% of the milk in New York came from swill dairies within the city itself. As a result of drinking swill milk, 'summer diarrhoea' – so named as it was more prevalent in warm weather – was rife. At one point in the mid-1800s, 1500 infant deaths a week in New York were associated with bottle feeding, including using swill milk.[4]

In a 2018 paper published in the *Journal of Food Protection*, researchers noted:

City milk was almost entirely distillery milk. These distilleries housed up to 2000 head of cows … The slop – slop it certainly resembled – was conveyed from large elevated tanks into wooden sluices leading to the cow stalls 3 feet wide. The cow consumed 32 gallons of slop

and 3 pounds of hay. Ulcers developed in the mouths of the
cows; their tails fell off. Tuberculosis of the glands, lungs,
and intestines followed. One Brooklyn distillery indicated
that, out of 1811 cows, 230 died in 10 weeks. Milk so obtained
was pale blue, often turbid, and malodorous. Peculating
[thieving] dairymen concealed their wickedness, not so
much as by diluting the milk with water, as by adulterating
the swill milk with plaster of Paris, charcoal, starch, sugar,
flour, and egg – making bad matters worse.[5]

This odious problem didn't go away quickly. By 1908, the US
Surgeon General still blamed the majority of childhood deaths on
impure milk.[6] It's hardly surprising. By that time, New York's milk
'came from 35,000 dairies situated in five different states, passed
through 400 processors and over 12 lines of transportation, and
was handled by 150 wholesalers and 12,000 retailers', according to
Richard Meckel in his 1998 book *Save the Babies*.[7] Lots of pairs of
hands, lots of transport – and this at a time before we had widespread
refrigeration.

While not the product of swill dairies, similar bottle-feeding
horrors were happening across the Atlantic in the 1800s. In Victorian
England, with the middle and upper classes rammed into corsets,
and breastfeeding difficult to do in such elaborate garb, bottle feeding
became fashionable. And the bottles themselves, with rousing
nationalistic names such as 'Empire', or evocative ones such as 'The
Little Cherub', were also made in fashionable new designs. The worst
was the 'banjo' shape, where a flattened bottle with a very wide base
narrowed to a tiny opening at the top where a tube attached – kind of
like a modern maple syrup bottle. This banjo shape made the bottles
almost impossible to clean properly, allowing dangerous bacteria to
proliferate. After the invention of intricate rubber tubes in the 1800s,
which attached even to easy-to-clean bottles, these dangerous baby
feeding systems eventually became known as 'murder bottles', as
infants fed through them started to die. According to the Baby Bottle
Museum in the UK (yes, there really is such a thing!), only one in

five infants who used these 'murder bottles' lived to see their second birthday.[8]

Adding to the problem was a misunderstanding of biology, and no understanding of germ theory. In her best-selling tome on cooking, cleaning and the hiring and firing of household staff (including wet nurses, as we saw earlier), *Mrs. Beeton's Book of Household Management*, Isabella Beeton suggested that the teats of the bottles only needed cleaning every 2–3 weeks.[9]

Between the dangers inherent in contaminated milk, and the way it was administered, it's incredible that people still drank it, or fed it to their children. But they did. Milk's reputation as a wonder food somehow overcame people's distrust, even as their neighbours and family members became ill.

By the time the Industrial Revolution had really taken hold, it's estimated that cow's milk contributed to the death of half a million infants in Britain between 1850 and 1950.[10] That's right, 500,000 babies and young children dead from the diseases and bacteria that could be carried in milk. In 1906, the risk of fatal diarrhoea in the UK was 48 times greater for infants fed fresh cow's milk, and 94 times greater for those fed condensed milk, than for babies who were exclusively breastfed.[11]

Now, although Louis Pasteur had shown how to make milk safer through pasteurisation in 1864 (more on that later, in the chapter The New Moonshine: Raw Milk), his technique took a while to catch on. In fact, the practice of feeding cow's milk and other animal milk to infants increased around the year 1900, as milk became easier to source, and cheaper. According to an article in the journal *Anthropology of Food*, the risks of unpasteurised, unrefrigerated milk from cows kept in crowded sheds included:

anthrax, botulism, brucellosis, cholera, diphtheria, dysentery, enteritis, *E. coli*, gastroenteritis, giardiasis, hepatitis, listeria, paratyphoid, salmonella, scarlet fever, tuberculosis, typhoid … hundreds of recorded milk-borne epidemics.[12]

The same researcher found that the middle class suffered disproportionately because of their higher milk consumption, and because their babies were most likely to be artificially fed. Further:

In about 1900, up to one third of samples of London milk contained pus from the diseased udders of country cows and in Manchester only 4.2 per cent of samples were found to be 'clean'.

The milk contained manure and dust from the cowsheds, along with dirt from the train carriages. Remember, this was a time before refrigeration – so all the pathogens in the contaminated milk had ample opportunity to flourish.

Milk, somehow, managed to maintain a wholesome image, and the Brits – like the Americans – kept buying it for themselves and their kids.

In 1865, Justus von Liebig invented something that would revolutionise baby feeding worldwide. As readers of my book *Soil* may recall, von Liebig was a chemist who demystified major chemical pathways in living soil, and also discovered the vital role of the major nutrients – nitrogen, phosphorus and potassium – for growing plants.

In humans, von Liebig went a step further, by concocting, patenting and then marketing the world's first baby formula. As one paper put it, 'Liebig's formula – consisting of cow's milk, wheat and malt flour, and potassium bicarbonate – was considered the perfect infant food.'[13]

It was called 'formula' because it was scientific, and required mathematical formulas to make. In theory, it was the magic bullet people had been looking for: a nutritionally complete replacement to human breastmilk. Liebig's formula was made into a dried form so it could be stored and transported more easily. Within 18 years there were 27 patents for infant formula, including one from Swiss company Nestlé, as the idea took off.

Formula feeding has been a boon for babies, and for mothers. No longer did we have to rely on the good nature (and good health) of a wet nurse. No longer did we need to worry about how clean a cow shed was, or if the animal that was milked was fed a diet of distilled grain that made them sick. Now we could feed babies how and when we liked.

Von Liebig's infant formula started a massive trend – yet his version was actually far from the perfect food for a newborn child. For a start, it was deliberately higher in protein than breastmilk. In 1867 a French obstetrician, Jean-Anne-Henri Depaul, tested von Liebig's 'food for infants' on four newborns, and all died of 'gastric distress'.[14] However, with time, and with tweaks, formula started to gain acceptance.

As science demystified milk's major nutrients, companies worked out how to mix more of the ingredients that a child could digest, in proportions that came closer and closer to breastmilk. This meant wet nurses were no longer required, and people could feed a newborn even if the mother was ill or had died. It meant that a mother who couldn't produce enough milk to feed her infant was suddenly given a safer, simpler option as a supplement – so long as they sterilised the equipment properly between feeds.

By the early 1900s, companies were muscling in on the mothering dollar. By that stage, one formula company had already boldly suggested that breastmilk wasn't good enough for tiny infants. Nestlé, which first patented its infant formula in 1878, encouraged women to use formula from 6–8 weeks of age.[15] In 1885, its advertisements stated that their 'Milk Food for Infants' was 'the only perfect supplement and substitute for mother's milk'.[16] In later advertisements, they offered free samples and suggested your baby would become anaemic if you didn't use formula from six months on. We now know, of course, that using formula short-circuits the feedback loop of infant-stimulated milk production in the breast – meaning the use of formula then becomes essential, not optional.

Even 30 years after the invention of both pasteurisation and factory-made formula, babies were still dying in massive numbers. Into this mess stepped Dr Luther Emmett Holt, an American who became the pre-eminent paediatrician and expert on all things baby. His seminal text from 1894, *The Care and Feeding of Children*,[17] has had 75 reprints.

In the book, Holt gives advice on everything from how to cure a baby who is tongue tied, to sterilising feeding equipment, and taking great care with cow's milk as a supplement. He wasn't big on night feeds, however, telling mothers to only feed a three-month-old baby once between 10 pm and 6 am, and to keep it that way until they were five months old. After that, he insisted, there were to be no night feeds at all. We now know this would likely lead to a loss of mother's milk, by disrupting the wonderful mechanism whereby a baby's drinking triggers more milk production.

Holt also provided a recipe for homemade formula, which includes milk, barley water and sugar. He also recommended that the occasional breastfeed was substituted with a bottle so the child gets used to it – because bottle feeding was considered equal, if not superior, to breast. Critics point out that Holt spent 15 pages of his book on artificial feeding of infants, and only four on breastfeeding (including advice to never feed for more than 20 minutes at a time), but perhaps that's ungenerous.

Formula feeding was on an upward trajectory, though not in a hurry. Mostly the great rise in use only happened after about 1950, partly thanks to a reticence to trust science, but mostly due to the prosperity boom that came in the wake of two world wars and the Great Depression. But when formula took off, it really took off.

According to one researcher, 'In 1944, 88% of Swedish mothers were breastfeeding their infants at 2 months of age; by 1970 the rate had declined to 30%.'[18] Around Kampala, the capital of Uganda, the proportion of children who were part bottle-fed before six months of age increased threefold, from 14% between 1950 and 1952, to 42% just a decade later.[19]

In 1940 in the US, 65% of babies were breastfed up until six months. By 1958, that number had dropped to just 25%, where it sat pretty much constantly until the early 1970s. It then rose again, to where the breastfeeding rate for infants under six months (including those also receiving solids or formula) now sits – about 51%.[20] In other words, about half the infants born in the US today are reliant on formula.

Australian numbers are pretty similar. While 96% of mothers start breastfeeding at birth, compared to about 83% of mothers in the US,[21] only 60% of Australian babies are still getting at least some breast-milk by six months of age.[22] By 12 months, only 28% of Australian babies are receiving any breastmilk compared to 35% in the US.

These numbers should probably be seen in the context of UN recommendations that babies receive *only* breastmilk for the first six months of life, and, if possible, complementary breastfeeding at least until they're 24 months old.

So, let's look at where formula is up to today, in terms of its macronutrient profile compared to other milks – and just what else is in it.

Milk nutritional comparison table[23]
Grams per 100 ml (3½ fl oz)

	Human breastmilk	Human colostrum	Infant formula	Cow milk*	Cow colostrum
Protein	0.8–0.9	2.5	1.8–3	3.3	14.9
Sugar	6.9–7.2	5.3	6	5.0	2.6
Fat	3–4	2.9	4.4–6	3.5	6.7

* Milk as it comes directly from cows, not from a carton

You may recall that von Liebig's formula had four ingredients – cow's milk, wheat and malt flour, and potassium bicarbonate. While modern infant formula is far more complex, it still usually contains some kind of modified cow's milk – over time, we've discovered that cow's milk has a lot of good things in it that are useful to a human newborn, so it still forms the base of most formulas today, albeit in a modified form.

Nestlé's Nan A2 Stage 1 Infant Formula lists the following ingredients:

Milk solids, vegetable oils, minerals (calcium citrate, potassium citrate, sodium citrate, magnesium chloride, potassium chloride, potassium phosphate, sodium chloride, ferrous sulphate, zinc sulphate, copper sulphate, manganese sulphate, potassium iodide, sodium selenate), emulsifier (soy lecithin), omega LCPUFAs (DHA from fish oil, AA), vitamins [sodium ascorbate (vit C), dl-alpha-tocopheryl acetate (vit E), nicotinamide (niacin), calcium pantothenate (pantothenic acid), thiamin mononitrate (vit B1), retinyl acetate (vit A), riboflavin (vit B2), pyridoxine hydrochloride (vit B6), folic acid, phylloquinone (vit K), biotin, cholecalciterol (vit D3), cyanocobalamine (vit B12)], choline bitatrate, inositol, taurine, L-valine, L-histidine, nucleotides (cytidine 5'-monophosphate, uridine 5'-monophosphate, adenosine 5'-monophosphate, guanosine 5'-monophosphate), antioxidants (mixed tocopherols concentrate, ascorbyl palmitate), L-carnitine, acidity regulator (citric acid), culture (Bifidus BL).

As nutritional science has improved, so has our understanding of what should be, and shouldn't be, in baby's milk. Nestlé's formula above has about 45 ingredients, though there could potentially be a few more with 'oils', plural, listed. The bulk of the mix is cow's milk (adjusted to be lower in lactose, among other modifications) – hence its appearance as the first ingredient on the label (as 'milk solids').

It also contains fish oil components, and usually an animal form of vitamin D (made from the oils found in sheep wool).

With advances in cell fermentation and nutritional science, there are now even vegan baby formulas on the market. In 2022, Nestlé also filed for a patent for a vegan version using potato protein in place of milk, though it was yet to go to market when this book went to print.

There is an infant formula for children over one year old made without milk, but far fewer that are suitable for a baby from birth. These are the ingredients of Sprout's vegan baby formula, which you can, apparently, feed to your baby from day one:

Organic rice starch, organic oil blend (organic coconut oil, organic canola oil), organic rice protein, organic pea protein, arachidonic acid (ARA), docosahexaenoic acid (DHA from algae), minerals (potassium, calcium, chloride, phosphorus, sodium, magnesium, iron, zinc, copper, iodine, selenium, manganese), vitamins (vitamin C, niacin, vitamin E, pantothenic acid, vitamin A, riboflavin, thiamine, vitamin B6, folic acid, vitamin K1, biotin, vitamin D2, vitamin B12).

That's 33 ingredients compared to Nestlé's 45. Both are trying to replicate what breastmilk contains naturally. (Note that vitamin D2 is the vegan form of vitamin D, which is harder for our bodies to use effectively, but seems to be the only non-animal-based option currently available.) Is it good for babies? We should hope so. It's hard to do studies on formula using anything except large population analysis and animal research, so we have to make educated guesses. In most countries, infant formula has very strictly controlled guidelines for how it is produced, and what it must contain, at least in broad terms. (In Australia, thanks to an international agreement on formula sales, there are even guidelines on how it is marketed online, with a requirement to ask that a person searching online understands that breastmilk is best, before they can click through to a formula maker's website.)

In the US, they have the *Infant Formula Act*, which came about after at least two infants died and over 140 infants became very ill from soy-based formula in the 1970s, prompting a review of all formulas and what they do – and must – contain.[24] When the first soy-based formula came out in 1929, it used the whole bean, and while many people reported problems with their baby's digestion (bad-smelling diarrhoea and the like), it kept babies alive. However, the isolated soy protein used in the soy formula made in the 1970s lacked vitamin K, with catastrophic results.

Should we trust formula? Yes, in places where there is decent regulation. Should we see it as a breastmilk substitute? Yes, but only if breastmilk (and importantly, breast *feeding*) is unavailable. Does it nourish a baby from day one? Yes. But is it the best option? No.

No mother should feel guilty for not being able to breastfeed her baby, for whatever reason, medical, mental or otherwise. Throughout history, we've always had trouble feeding babies, and while breast-feeding *is* optimal, what is optimal isn't always possible, in everything in life. Using formula might be less good, but remember, it's still good. Any nourishment is important for a child. And besides breast-milk, formula is the most useful thing in the toolkit so far.

Race has also played a part in formula's rise. In her book *Skimmed: Breastfeeding, Race, and Injustice*, American law professor Andrea Freeman relates a story about the first identical quadruplets to be born alive in the US.[25] Born to a Black–Cherokee mother, Annie Mae Fultz, in 1946, the four daughters were thrust into fame just as they came wriggling into the world. So much so that the white doctor who delivered them, Fred Klenner, saw not just babies, but *opportunity*. An unabashed white supremacist, he firstly appropriated the role of naming them (calling them all Mary, and with their second names coming from his sister, wife, aunt and great-aunt) – then also auctioned off the rights to use the girls to advertise infant formula. Annie Mae, who'd lost the ability to hear or speak after a childhood illness, had little power to stop her children being used to market formula *en masse* to Black Americans, who until that time were reticent to spend money on more expensive formula and preferred to breastfeed.

PET, the formula company that won the rights to use the girls to advertise their baby formula, found their profits soared to record levels.

Milk and murder

Formula is no stranger to controversy. From dubious practices such as giving away free samples (which effectively helps to dry up a breastfeeding mother's milk), predatory marketing using clandestine techniques in developing countries, all the way to deaths from formula that was poorly produced or even tampered with, it has a chequered history.

By the late 1930s, a paediatrician named Cicely Williams had seen enough of the practices of companies selling infant milk substitutes to poor women in Malaysia to make her angry. Very angry. A leading doctor at a time when very few women could access the profession, Williams was the first to diagnose kwashiorkor, a wasting disease caused by protein deficiency. Having worked with the poor from 1936, in 1939 she was invited to address the Singapore Rotary Club, whose chairman just happened to be the president of Nestlé – a company implicated in quite a few infant milk scandals over the years.

What Williams had seen made her livid. Despite it being illegal to promote and sell condensed milk to feed infants in the UK and Europe, multinational companies were doing just that in Malaysia, and other developing nations. She witnessed company representatives dressed as nurses, recommending artificial feeding for babies, using sweetened condensed milk.

Incensed, Williams titled her talk 'Milk and Murder', and didn't hold back, saying:

> If your lives were as embittered as mine, seeing day after
> day this massacre of innocents by unsuitable feeding, then
> I believe you would feel, as I do, that misguided propaganda
> on infant feeding should be punished as the most criminal
> form of sedition, and that these deaths should be regarded
> as murder.

Williams noted what many later confirmed: predatory practices by international firms to increase sales.

Despite being a feted doctor, one of the first women to practise medicine in the UK and being well respected in her profession, Williams' plea didn't fix the problem. Nearly two decades later, in 1956, colonial publications sanctioned by the British government were still doing damage. For instance, in *Gold Coast Nutrition and Cookery*, a UK-funded book in what is now Ghana, the authors heavily promoted breastmilk substitutes and artificial feeding, explicitly stating:

> Breastmilk, even during the first 6 months of a baby's life, does
> not completely supply his needs, and must be supplemented by
> other foods.[26]

While regulation of baby formula continued to improve in richer countries, it wasn't until the 1970s that a charity and advocacy group called War on Want brought the spotlight to what was going on in developing nations. In their booklet *The Baby Killer*,[27] published in early 1974, they highlighted the rapacious practices of formula companies, including the provision of free samples and false or misleading advertising. When it was translated into German, the pamphlet was titled *Nestlé Kills Babies*. Unsurprisingly, Nestlé sued, and perhaps more surprisingly won – though not with any glory, according to reports from that time.

The Baby Killer detailed research that showed how mothers stopped breastfeeding after receiving free samples. They showed how mothers in poorer nations were encouraged by formula company representatives to try milk substitutes, just in case they didn't produce enough breastmilk themselves (and we know most women can usually produce enough if they eat enough themselves). The authors also pointed to baby formula's heavy economic burden. In Egypt and Pakistan, formula feeding a three-month-old infant cost 40% of the average monthly wage. Even worse, the babies were getting sick, partly because of the cost, and hence over-dilution of the formula, and partly because many households didn't have the gear or knowledge to sterilise bottles properly.

The Baby Killer and other research prompted a massive international public boycott of infant formula in 1977, and also triggered a global agreement, the *International Code of Marketing of Breast-Milk Substitutes*, which came about in 1981, after seven years of consultation. When it was passed by the United Nations, 118 countries voted in favour, and only one country voted against the code – the United States.

That code is why, when I look up an infant formula in Australia, there are rules about how it is sold, and all information must be prefaced with an acknowledgement that breast is best. You can't advertise formula, or the paraphernalia to use it (bottles, teats) to the general public, or give away free samples. You can't give formula to maternity wards or infant care centres. You can't try to undermine breastmilk, essentially. Sadly, however, because it's an international code, it relies on national governments to police. At last count, only 39 out of 194 countries studied had legislation that fully reflected the code.[28] And globally there are 500 or so reported violations each year.

The code makes things better, but not perfect. According to a 2023 report in the *British Medical Journal*, looking at what formula companies say their products do, and what their products actually do:

most products did not provide scientific references to support claims, and referenced claims were not supported by robust clinical trial evidence.[29]

The more things change, the more they stay the same.

What happens when formula fails?

In 2022, there was a crisis in baby formula in the US. There were production delays because of Covid – and then there was a recall of major brands Similac, Alimentum and EleCare, made by a single company called Abbott, after harmful bacteria were discovered in the powder. Two babies apparently died and Abbott's was likely the cause, while more deaths and illnesses were implicated.

The mass withdrawal of formula from shops and a closure of the production plant meant there wasn't enough formula to go around. Given that Abbott produces 40% of America's formula, and most babies in the US rely on supplementary feeding in the first six months of life, people started to panic. Baby formula was being stolen from shops, and also sold on the black market. Prices rose 18% in a year, with some American states only having 10% of the normal supply. Ripples from the crisis spread around the world. The cheapest brand in the UK rose 45% in price in two years. People were watering down their formula using cow's milk – or worse, condensed milk. Or just water.

In early 2022, the Biden administration in the US, in a program called Fly Formula, actually fast-tracked shipments and chartered planes on 32 different flights to bring in 19 million bottles of formula from overseas.[30] That same year they also passed the *Access to Baby Formula Act*[31] to allow more access for disadvantaged groups in particular, as well as more flexibility for the regulator to look at removing the dominance of major players and their ability to disrupt supply.

As we saw earlier with the soy formula deaths in the 1970s, relying on a nutritionally unbalanced or poorly made formula can have terrible results. In 2003, three babies in Israel died and 23 were seriously harmed after a German formula company, Remedia, sold a vegan soy powder that was found to be lacking vitamin B1 (thiamine) – a nutrient that exists in soybeans, but not the heat-treated extract they were using.[32] A few Remedia executives faced manslaughter charges, but only one, Frederick Black, Remedia's chief food technologist, was convicted in 2013 in an Israeli court.[33]

Even this horrific outcome looks relatively minor compared to the Chinese formula disaster of 2008, where at least six infants died, and a whopping 300,000 became sick after formula companies put melamine in their mix. Melamine, a building material used in plastic, boosts the protein score in milk tests, but is dangerous to health – especially when it was found in formula at levels 3000 times higher than permitted in the US. In all, 22 companies were caught up in the mess.

The Chinese formula scandal was a result of the nation's lax policing of standards. Even today, Chinese people distrust their local brands – with flow-on effects reaching as far afield as Australian supermarkets, which often sell out of baby formula when 'personal shoppers' strip the shelves of tins and ship them back to China. People also carry tins of formula back to China in their luggage, it's so prized. Australian shops have resorted to putting limits on how many tins you can buy, as well as using unremovable locks on the top of the tins to prevent theft.

Today, China has some of the strictest rules of any country around formula manufacture. And they need them. In the early 2010s, in the western and central regions of China, only 58.3% of mothers exclusively breastfed their newborns, and only 13.6% did so at 5–6 months. And at six months, the age at which all babies ideally would exclusively still be on their mother's milk, the national rate of *any* breastfeeding was only 20.8% – one of the world's lowest.[34] The reasons so few women breastfeed are complex, but much of the blame is attributed to the way formula is aggressively marketed.

Yes, still.

All healthy animals that lactate produce a milk designed for their young. It's unique not just to their species, but to the individuals who produce it, and drink it.

Around the world, it's estimated that each year, 823,000 child deaths could be prevented through appropriate breastfeeding – as well as 20,000 maternal deaths from breast cancer.[35]

Comparing formula to real milk is like comparing a pine plantation to a rainforest. Complexity in nature is something we humans have usually underestimated. What seems good (or 'perfect', as was the case with von Liebig) in one era is quickly discovered to be lacking. If there's one consistency over the centuries, it's that what we *think* we know is either wrong, or not nuanced enough. Milk has astounded us with its properties since we first had conscious thought. It has proved essential, magical and, at its heart, probably irreplaceable, despite our hubris in thinking we can improve on nature.

Infant formula is vital to keep many vulnerable babies alive. But it's also an industry, estimated to be worth about US$70.6 billion in 2019. The problem comes at the intersection of formula and capitalism, where whole cultures – such as the Chinese, whose market share is nearly a third of the global total[36] – can be subjected to influence that suggests formula is better than the real thing.

Developed nations aren't immune to this coercive behaviour, either, according to the *Archives of Disease in Childhood*:

In the UK, where the rate of exclusive breastfeeding (7% at 4 months) is one of the lowest in the world, companies spend 10 times more on advertising than the Department of Health spends on promoting breastfeeding.

And the reason?

20% of mothers in the UK who were weaning their babies at 4–6 months of age thought formula was better and more nutritious than breastmilk.[37]

What formula does is keep babies alive, and eventually allows them to thrive. But it isn't a perfect replacement. Real milk contains growth factors, sugars that a baby can't digest, fats that alter the baby's brain development, genetic codes, and enzymes whose role we are only just beginning to comprehend.

Later, I'll look at how milk is formed in the body, and do a deep dive into what real milk does, and what formula can't yet – and may never – achieve. Trust me, you'll never look at real milk in the same way again.

As we saw earlier, milk plays a vital role in immunity – something that has implications for us as individuals, and possibly humanity as a whole.

And to understand that, I need to talk to a doctor about poo.

Don't Mind
the Backwash

Daniella has a freezer full of pregnant women's poo.

I guess that's the price you pay for being an obstetrician and gynaecologist obsessed with how to get the best outcomes for at-risk pregnant mums. Dr Daniella Susic, a senior lecturer with the University of New South Wales, is one of those vital people who looks at everything from the father's diet pre-conception through to the mother's microbiome (which she samples from their mouth, vagina and poo), to the role of milk in the gut health and immune response of babies.

On our first meeting, when I originally learned about her oddly filled freezer, Daniella became interested in pigs. I was telling her about research that shows a pig's gut health, and hence their immune status, is directly related to their exposure to soil[1] – which had led me to start digging into the benefits of their milk.

Daniella's interest in human health parallels work done on farm animals, and while she's worked with female baboons, she's never worked with another biologically similar species to us, the pig.

Daniella, however, is more interested in humans. And, so, I went to her cottage to talk milk. And to talk milk, we had to talk mums, boobs and babies.

A pig is a good case study for Daniella because it's born with very little immunity. We can also do way more studies on pigs than people, for ethical reasons among others. Compared with humans, who some think can get 95% of their initial immunity from their mother's placenta and a vaginal birth, pigs gain 95% of their initial immunity from their mother's colostrum.[2] That initial milk a piglet drinks is packed full of protein, about 16% in total. And while there is some milk protein in there (casein and whey proteins), most of the protein in pig colostrum is made up of antibodies.

Antibodies are the magical immunoglobulins – think of them as pathogen fighters – that protect an animal when unwanted things (diseases, viruses, bad bacteria and so on) enter its body. If piglets don't get enough colostrum in their first day of life, 50% of them die, regardless of milk feeding afterwards.[3] Their newborn gut is very leaky, allowing the antibodies – essentially the mother's own immune library – to inoculate the piglet's body quickly and efficiently.

When a baby pig first drinks, a whopping 70% of the proteins in the colostrum are likely to be these disease-fighting antibodies. However, only six hours after birth, a piglet's initially porous gut starts to close to the benefits of colostrum, and therefore to these antibodies. The antibodies drastically drop in number – some by up to 90% within 24 hours.

On our farm, we've seen the effect in our own baby pigs and goats, whose growth remains stunted if they don't get the magical colostrum from their mothers within the first 24 hours. If they survive at all.

This seems to be mirrored in human babies. Daniella tells me that in humans, lots of immunoglobulins are passed to the baby through the placenta. At birth, however, that tap is obviously turned off, and the conferred immunity then switches to breastmilk. 'Boobs are the best thing ever,' she says. 'These antibodies in breastmilk are the baby's first line of defence.'

So, getting a baby to drink colostrum early is vital to its health.

But then, so is their access to milk after that first day. Let's look at oligosaccharides as an example of just one component of milk that has long been underestimated, and which ramps up in milk after the first day's colostrum. These complicated-sounding things in a mum's milk are sugars that babies, and piglets, can't digest. In fact, there are over 90 of these indigestible milk sugars in sow milk, and over 300 of them (that we know about) in human milk.[4]

If we can't digest oligosaccharides, why on earth are they there? Science has only recently worked it out.[5] 'The only reason for their existence is the priming of the gut,' Daniella tells me. Their main job is to set up our guts to breed bacteria. Let me explain.

The connection between gut health, human health and our microbiome is all quite a recent discovery to science. Only a couple of decades ago, we used to think that milk as it came from a healthy animal, human or otherwise, was pretty sterile. That is, there were no bacteria, or at least very few, in milk. When we found bacteria, we presumed it was a handling or processing error; things that had slipped in when the milk was collected. Recent research has found at least 220 species of bacteria that exist in sow milk.[6]

So, what about human milk? Is it teaming with bugs? We now know that if a baby drinks 800 ml (27 fl oz) of breastmilk in a day – which is about average for an infant up to six months – they are also ingesting up to 8 million commensal (good) bacteria from about 820 species each day.[7] These 'bugs', once called 'germs', are there to protect the child. (Commensal bacteria act by triggering the baby's immune system, which prevents or hinders the breeding and action of more dangerous bacteria and other pathogens.)[8]

We are not alone in our bodies, and that's now seen as a good thing. Over half the cells in our bodies – about 56% – aren't ours. Those other non-human cells are mostly bacteria, as well as archaea, viruses and fungi – all of which have a positive function in the right numbers and in the right place. Those bacteria, some 39 trillion of them (that's 39 with 12 zeros), help us digest the food in our lower gut, for instance, by fermentation. But they are also intimately tied to all sorts of other processes, including immunity and brain health.

Many researchers are now calling the gut, the hotspot of these bacteria, our second brain, because the brain and gut are intrinsically linked by physical and chemical connections known as the gut–brain axis. What happens in our gut can affect our mood, our ability to think. Our brain health.

The bacteria and other organisms that live in and on us are called our microbiome, meaning our microscopic life. And the microbiome in mammals is intimately tied to their milk.

Just 20 years ago, an aeon in milk research terms, we knew that pig milk contains substances that are involved in 'stimulating gastrointestinal tissue growth' to ensure the gastrointestinal tract matures, while also helping to repair any damage to the lining of the gut. Substances the scientists describe as 'milk-borne bioactive compounds' include things such as 'immunoglobulins, lactoferrin, lysozymes, lactoperoxidase, leukocytes, epidermal growth factor, insulin-like growth factors, and transforming growth factors'.[9] We'll meet some of these again later, but essentially, all you need to remember is that there's a bunch of stuff in milk that isn't there by accident. It's a warehouse of biologically active and functionally important ingredients.

One of the most vital are those oligosaccharides we met just before – the milk sugars that are indigestible to most mammals, including baby mammals. For a long while we thought of oligosaccharides as surplus to our needs. Kind of like the way we once thought we had 'junk' DNA, when we couldn't work out what all our non-active genes did.

It turns out, of course, that oligosaccharides aren't junk, and they *are* digestible – just not by us. Rather, they're consumed by our gut bacteria. These milk sugars, the ones that Dr Daniella loves so much, actually have at least two vital functions: to feed our beneficial bacteria, and to stop dangerous bugs attaching to our gut wall. In pigs (where happily for us, and sadly for pigs, we can do more interventionist research), oligosaccharides have been shown to breed up good bacteria, and also drastically increase the formation of certain cells, called enterocytes, that line the gut. Enterocytes are the cells that let us absorb nutrients from food in our small intestine. The 'junk' sugars in milk therefore have an active role in maturing a baby's digestive system.

In mice, where even more research has been done (at great cost to mice, it must be said), the rate of enterocyte cell formation is increased by up to 80% thanks to the milk sugars we once thought of as rubbish.

As we dig deeper and deeper into the science of milk, we're constantly learning more. For instance, human milk oligosaccharides actually boost the gene response of our immune system, while increasing the ability of cells to produce antibodies.[10] They trigger a genetic fast-tracking of immunity, which has immediate and measurable impacts on a child's ability to fight disease. Milk oligosaccharides do this without even needing to feed the gut microbes, though exactly how they do this is yet to be elucidated. When we drink milk as adults, cow milk oligosaccharides are still feeding our internal gut flora. Not bad for something we once thought was filler.

Milk is part of a biological system. A mother's body is also involved in a very complex biological dance that means a mother's milk is actually designed very much with the individual baby's needs in mind. Everything from smelling a baby's head to a kiss on its cheek to the act of suckling allows a mother to actually personalise her milk in real time, and change what is in it. And some of this has a strong effect on the child's immune system.

Before we get to that, let's look at milk's innate immune-boosting abilities. Its magic properties, way beyond its obvious nutritional density.

Milk, of any kind, has an immediate immune benefit on the drinker. It coats the mouth and the entrance to the nasal cavities, and lines the top of the windpipe – preventing dangerous bugs sticking to those sites. The microbiome of raw milk – the stuff we drink in infancy from our mothers and some drink from their own dairy animals – includes beneficial lactobacillus bacteria, which outcompete bad bacteria from the mouth down. Many of the bacteria in milk form the basis for the child's own microbiome, their gut flora, which they will carry throughout life.

But milk's role in our immunity is far more complex than it first appears. For a start, milk is a mixture of incredibly nutritious bio-active compounds, and it reacts to a very complex living being: the milk drinker.

After lining the mouth, the second thing a mother's milk does is react with the infant's saliva. This simple mixing has some only recently discovered benefits. Compared to an adult's, baby spit has 10 times more of two vital chemicals, hypoxanthine and xanthine – which, when mixed with human breastmilk, create hydrogen peroxide.[11] So much hydrogen peroxide that the combined effect of baby saliva and the antimicrobial properties of breastmilk actively kills real baddies such as staph and salmonella.

The saliva also helps breed up all the good bacteria conferred by breastmilk, which then inoculate the baby's gut, helping to boost their general immunity, from the mouth all the way to the gut.

There's some contention about the origins of the bacteria that exist in a mother's milk. Some suggest they come from the surface of the skin – that they get into the milk as the milk comes out. Others say they're picked up from the baby's mouth and get into the milk that way. And there's evidence that bacteria in the mother's gut are somehow transported internally through the mother's body and into her milk (though we don't yet know how).[12] Is it a mixture of all three?

Even more remarkably, a mother's kiss to a baby's head can alter the makeup of her breastmilk.[13] Scientists have long wondered if some innate maternal behaviours have biological benefits, as well as the more obvious psychological ones. For instance, humans aren't the only species to press their mouths against their young. Goats do it. Sheep do it. Dogs do it. Primates all do it. Is there something more to this beautiful and very simple act?

Imagine the baby has some kind of virus or other pathogen on its skin. A newborn baby's immune system is too immature to fight many diseases – but the mother can help. When a mum kisses her young baby, some of the virus enters the mother's mouth. If she's had the virus before, the mother's body recognises it instantly and accesses her immune cells that recognise the virus and can fight it off. These immune cells rapidly multiply in her lymph system and pass into the mammary glands, where they are laid down in her milk. It happens in barely hours. Hence a cocktail of ready-made immunity, tailored to the child's needs that day, is delivered via the breast at the very next feed. Incredible, hey?

Immunity from antibodies in breastmilk is called *conferred* immunity. It's a gift from one generation to the next. Kissing a baby's head gives joy to the mother, and immunity to the child, in a majestic feedback loop.

There's another biological circuit that science is still trying to get its collective head around: the ability of a baby's saliva to create a direct response in the mother's mammary glands, without even the need for a kiss to the head. And it's all about the backwash.

In a scientific paper in *Nature* in 2022,[14] a team of researchers talk about 'retrograde duct flow' and its role in conferred immunity.

This is really early research (only a few months old at the time of writing). It's only been confirmed in mice, but it's likely that it happens in most mammals, including humans – and it's utterly remarkable.

What happens is this. If a baby has been exposed to a bug – a rotavirus, for instance, which is the most common viral cause of diarrhoea in infants – their saliva will contain the virus. When the baby feeds, it creates suction on the mother's nipple, which in turn creates a bit of a vacuum. When the baby unlatches, saliva from the baby is sucked back into one of the several holes in the mum's teat, into the milk ducts and into her mammary gland. Now, so far, it sounds dangerous.

But mothers have been exposed to a lot of pathogens over their time on Earth, and their immune system is way, way more mature than their child's. When the virus enters the mother this way, her immune response to that virus is way quicker than if she herself was infected. She quickly mounts a response that is targeted to her breastmilk, and her mammary glands provide antibodies specifically for her baby.

It's yet another elegant biological mechanism, using the mother's library of immune responses to fight threats in the infant – but much more quickly than we saw with the kiss.

These things sound incredible, but the more we look at the natural way of things, the more we see there is a system that has evolved to keep babies safe. Milk feeding might look simple at first, but the more you discover, the more miraculous and complex it becomes.

For instance, we're still learning just how much better actual breastmilk is for our young than infant formula. In some novel research, again using pigs, researchers found there were four times the antibodies in the blood of piglets fed human breastmilk than those fed formula milk. That's a whacking great increase in general immunity. They also found those on breastmilk had more cells that could mount a specific immune response to a specific disease they tested – in this case, cholera. In fact, in those fed breastmilk, these immune cells were 13 times higher in the lymph nodes, and 11 times higher in the gut. At around seven weeks old, the breastmilk-fed piglets also boasted more T cells, a type of white blood cell that is intimately tied to immunity.[15]

Now, when we talk immunity, and nutrient-dense food, milk is the bomb. And it's not just mother's milk that counts. Milk of just about any kind has some unexpected beneficial impacts. Along with the bacteria, oligosaccharides and the antibodies we met earlier, there are also lots of other biochemicals in milk. Things such as lysozyme (which is found at about 1000 times the concentration in human milk than in cow's milk). There's lipase, which helps break down and digest certain fats and can play a role in liver function. There's osteopontin (again, 10 times more concentrated in human milk than in cow's milk) – a chemical that has important immunological properties, and also has a role in collagen formation. (Collagen isn't just the beef-sourced form that is injected into people who want fatter lips or fewer smile lines. It's actually a vital connective tissue in skin, tendon, bone and cartilage.) Osteopontin is good for all your tissues, particularly for wound healing, while also helping inhibit cancerous cell growth.

Let's not forget alpha-lactalbumin, the second most common protein in cow's milk whey. While it improves the absorption of iron, it's also a vital secondary enzyme for digesting lactose. And, as with many of these biochemicals, its role is way more complex the more we look into it. For instance, it provides essential amino acids for optimum growth in the newborn, and the peptides (protein fragments) it releases during digestion act as antimicrobial agents. And while fighting baddies is good, helping goodies is even better. Of course, alpha-lactalbumin does that too, like an annoyingly talented friend who's good at everything. It increases the growth of bifidobacteria, a family of beneficial bacteria that is used in yoghurts and is associated with gut health.

Alpha-lactalbumin also contains lots of tryptophan, a precursor to serotonin, one of the happy hormones that gets released in our brains. Apparently, in stress-vulnerable adults, supplementation with alpha-lactalbumin can enhance brain tryptophan and sero-tonin levels, which are associated with improved cognition, and better memory and sleep.[16]

Whoa! So alpha-lactalbumin helps anxious people think clearer, remember more and sleep better. Sounds too good to be true.

But wait, there's more. Milk also has cytokines, which are crucial for the healthy functioning of immune cells and blood cells more broadly. They control the growth and activity of immune cells, which has implications for how our bodies ward off cancer.[17]

All this, and I haven't even mentioned lactoferrin, which you don't find in plant-based foods. According to a 2019 paper in the *International Journal of Food Properties*,[18] lactoferrin has:

> **beneficial properties such as anti-pathogenic, anti-cancer, anti-inflammatory, immunomodulatory and DNA-regulatory activities. Recent reports indicate its therapeutic properties in the treatment of neurodegenerative diseases associated with aging, as well as stress-related emotional disorders.**

Because it absorbs iron easily, lactoferrin helps prevent the proliferation of bad bacteria, which are usually more iron dependent. It also:

> **endorses bone growth, protects the intestinal epithelium, and stimulates the retrieval of immune system functions in animals. Lactoferrin is involved in the treatment of hepatitis C. It also improves the health status of Alzheimer's disease patients.**

Wow.

And while lactoferrin exists in cow's milk, it's about 20 times more concentrated in human breastmilk.[19] Camel milk is also a cracking source – which, incidentally, along with other constituents that reduce the ability of bad bugs to breed, means it also keeps for longer than cow's milk out of the fridge.[20]

Of course, whenever we discover something as miraculous for our bodies as lactoferrin, there are those who would like to synthesise it and just add it to food. To put it in things where it may, or may not, do some good. They're already producing a genetically modified rice containing the human breastmilk gene for lactoferrin.[21] It's probably good to put lactoferrin and some other important factors in infant formula, things that a baby would otherwise have received through breastmilk. But if we've learned anything from history, it's that what we add to processed food isn't usually absorbed or used by our bodies in the same way as when it is consumed in real food in a balanced diet. Lactoferrin, for instance, we know is far better at doing its job if it's around well-solubilised calcium – as it is in milk.

This complex, immune-boosting, nutritious substance we drink in early life, and sometimes throughout life, is the result of many things. But mostly it's the end result of a lot of evolution, where what is in the milk of an animal is intimately tied to its infant's needs (and can fulfil a human's needs, too).

How it is made is, like all biological systems, a thing of wonder and amazement, something we humans are only just beginning to fathom. That half the human population is walking around with the tools to produce this miracle product makes it no less extraordinary.

Come with me as I take a look at how milk is made.

How Real Milk is Made

In 1885, a bloke named William Hay Caldwell killed, or had others kill, over 1300 echidnas in the name of science.[1]

Why? In the mid-1800s, white colonialists in Australia didn't really know what they were encountering when they looked at our wildlife – especially monotremes. Caldwell had just 'proven' to Britain's Royal Society that the platypus was an egg-laying lactator (because the Brits didn't believe Aboriginal people, who'd known that for a good 60,000 years). With his new-found fame, Caldwell became keen on documenting the entire egg-laying, milk-producing life cycle of another monotreme, the echidna. To do that, with the Society's blessing, he simply knocked hundreds of them on the head.

Learning about milk has meant that we've had to learn about us, about things similar to us, and about mammals unlike us. Every time we consider milk, what we drink as babies and as adults, there's usually a lesson in nature. Though hopefully with less of the slaughter.

For a long time, Europeans in Australia didn't consider that the platypus could be an egg-layer, and at the same time possess mammary glands. Perhaps that's because a platypus lacks teats. In the absence of nipples, it was assumed that the babies simply hatched from eggs. But when other early colonists did autopsies on the platypuses and found what were undeniably mammary glands, it was decided that, yes, platypuses must lactate – but not lay eggs. Much of this intransigence was in response to evolutionary theory. People were trying to reconcile differences between different species, while at the same time arguing that God's line of what-was-what couldn't be crossed: birds, fish and reptiles lay eggs; a milk-producing mammal doesn't.[2] (In the same way, conservatives of the era believed that people of different classes couldn't cross into another class.)

It wasn't until 1884, thanks to Caldwell's evidence, that the British scientific community was satisfied that platypuses did both. And if you're wondering what became of those 1300 clobbered echidnas, naturalist Jack Ashby in his book *Platypus Matters* recounts that Caldwell took his information back to the UK, submitted a few preliminary findings – then promptly published nothing more about them. Not a great end for the echidnas, but a blessing for Australia's platypus population, which was next on Caldwell's list of things to research even more deeply.

We now know that platypus milk is one of those miracle products that kills superbugs. The whole animal is a miracle, really. When in full lactation, the mother can eat her entire bodyweight in a day, just to produce enough milk for her two 'platypups', as Jack Ashby calls them. She has to swim kilometres – potentially over 10 km (6 miles) a day, covering up to 58 hectares over three months – in search of invertebrates for food. That's how draining lactation can be.

There were always clues to the platypus's unusual reproductive strategy. The young have an egg tooth, nature's built-in hammer to allow them to escape their shell, but unlike birds and reptiles, this tooth stays intact for months. Why? Because when the platypup rubs its mother's stomach with their rubbery version, it helps trigger milk letdown. A platypus doesn't have nipples, which would be pretty awkward if your babies have a duck-shaped bill for a mouth – so instead they lap milk from the mother's fur. Like marsupials more generally, platypus babies are altricial, meaning barely formed, and need nutrient-dense milk to finish developing. Exuding a nutritious milky substance from the mother's belly when rubbed sounds exactly the ticket for this exact animal.

So, too, is the flattened, elongated shape of the monotreme's mammary glands. A platypus's two mammary glands lie against her chest, just slightly back from her middle. Together, they can take up about a third of her underside – pretty much her entire belly from front to back legs. They can fill so much with milk that they start to stretch out over her back. If you have to spend hours each day swimming in murky, rocky streams with submerged branches, protruding lactating boobs wouldn't be a good thing.

Mammary glands are a thing of wonder. What they produce, how they produce it, and the variation in them among different species is astonishing. There's blue whale mammary glands, weighing up to a quarter of a tonne. Dwarf mongooses that have never been pregnant can spontaneously lactate to help rear the young of other mums. High-yielding dairy cows might have 50 kg (110 lb) of mammary gland hanging below their body when they go into the milking shed – less than half of that will be the actual milk; the rest is connective tissue and fat. In a human, the mammary glands are held more generally in the breasts, which can weigh 1 kg (2 lb) each. Yet an elephant, despite its size, boasts two mammary glands each only about the size of a rockmelon (cantaloupe), producing about 11 litres (2½ gallons) of milk in a day from her 150 kg (330 lb) of food. Meanwhile, if you're a lactating mother and know how demanding it can be to suckle an infant, spare a thought for dolphins, which have to suckle their young every 20 minutes, 24 hours a day, when they first give birth to a calf. And some dolphins nurse their young for 4½ years.[3]

Mammals, of course, take their name from the fact they produce milk. But the name actually derives from *mamma*, from the Latin for 'breast'. And, of course, *maman*, *mum*, *mom*, *mamá* and *mamma* are all words used for mothers today, the connection to mammary glands unmistakable.

So how is milk actually made?

Mammary glands are formed in the embryonic mammal, in humans starting just six weeks after the egg is fertilised. In most mammals, they remain relatively unchanged until pregnancy, but human glands enlarge at puberty. During pregnancy, the hormones prolactin and placental lactogen are released, which increases the breast gland tissue of the mother, but also specialises its function to enable the tissue to produce milk. Despite differences between species, from monotremes to primates, there are some major similarities between all mammary glands.

Some mammals have two of these glands, others have over two dozen. In humans they are situated on the pectoral muscles, essentially under the armpits. In horses they're on the abdomen. In pigs they run in pairs from the chest to the lower abdomen. And several species of the African rat genus *Mastomys* – the so-called multi-mammates (meaning 'many breastfed') – have up to 12 pairs of mammary glands, whereas other mice only have five. To fit them all in, mastomys nipples run all the way from the armpits along the bottom edges of the body to inside the thighs.[4]

The most complex mammary glands in the animal kingdom are, reportedly, those of the red kangaroo,[5] which can suckle two joeys at once – on markedly different milks, with one being newborn and the other joey already out of the pouch. They have four mammary glands, though strangely, two are never used.

In humans, as in many mammals, much of the structure of the breast may be fat or connective tissue. Size of an empty gland is no indicator of milk production in a cow, just as it isn't in a woman.

Drawings of mammary glands tend to look like images of the Nile Delta. Or lace. Or a head of broccoli. The heads of broccoli are actually about 15–20 milk-producing lobes in a human breast. A series of tubes coalesce from those lobes towards the teat, or nipple. Branches come from fine ducts in the broccoli florets further out, to a more central, fatter duct, or 10 further in. It's in specialised glands in the florets, which scientists think evolved from a type of sweat gland, that milk is made. Milk is constructed from what is provided to those glands, both from the blood, and from the lymphatic system.

The lymphatic system runs in tandem with our bloodstream, keeping the fluid content of blood at desirable levels. It is also very intimately tied to our immunity, ferrying pathogen-fighting cells from bone marrow and blood to where they're needed. Lymph nodes swell when you get an infection or disease, as they build immune cells and fight bacteria and dump the dead ones in the nodes. There are hundreds of these nodes all over the body, and you might feel the ones under your arms or in your neck swell if you're sick. Lymph's role in the mammary gland is smaller than that of blood, but just as mighty.

The mammary gland has lymph and blood vessels coursing all around, because the transfer of what is needed in milk from the mother's body can only happen through porous membranes at the microscopic level. Given that the richest source of nutrients to the mammary glands is the mother's blood, a dense blood supply to those glands is pretty vital.

In a cow, to produce a litre of milk, 500 litres (110 gallons) of blood must pass through her udder.[6] So when my cow gives me 10 litres (2 gallons) of milk a day, plus nearly the same amount for her calf, that's 10,000 litres (2200 gallons) of blood that she needs to circulate to her udder in a day. These numbers are similar to those in mother pigs, even though the blood supply system to their glands is completely different.[7] In women, the amount of blood supply to the breasts varies a lot between individuals, but at least one study shows that the left breast typically gets more blood supply than the right.[8] And overall, it matches the 500:1 ratio in terms of blood to milk seen in other mammals.[9]

It sounds like a lot. Luckily the human heart pumps about 100,000 times a day, and in the process moves about 24,000 litres (5300 gallons) of blood over about 19,000 km (nearly 12,000 miles) through all your arteries, veins and capillaries.[10] To produce about

800 ml (1½ pints) of milk in a day, a woman only needs to shift an extra 400 litres (90 gallons) of blood, so it's not an impossible ask, just a 2–3% increase. But it's still an extra biological function that has to be performed, raising the mother's metabolism, which usually also increases her appetite – though not quite as much as a platypus's.

Once in the mammary gland, blood is mined for its contents. Blood carries all the body's nutrients around, so it's essentially a jam-packed smorgasbord of ingredients from which the body can make milk.

In olden times, milk itself was known as 'white blood', being such a mysteriously vital white elixir that came from the body and nourished a baby with rich life-giving nutrition. So nutritious, even for adults, that Christian orders such as Roman Catholicism considered milk in general a form of 'meat', banishing its consumption on Fridays, as well as fasting periods such as Lent, the 40 days before Easter.

Milk isn't, as those forebears imagined, a form of blood, but it is fashioned from the blood supply. The mammary glands contain tiny cavities called alveoli – similar in function to the alveoli we have in our lungs. In our lungs they trap oxygen from the air we breathe and transport it into our blood. In the mammary glands, the alveoli are lined with milk-producing cells called lactocytes. ('Lactocyte', rather charmingly, means 'milk jar'). Cells in the alveoli draw the ingredients needed for milk (such as phosphorus and other minerals, some vitamins, fats, carbohydrates and protein) from the blood, and the lactocytes make milk proteins such as lactalbumin and casein, micronutrients, hormones, fats and lactose, along with cytokines, which regulate the growth and effect of immune cells. Groups of these alveoli form lobes of the mammary gland – the broccoli florets in the analogy I used earlier.

The lymph system's role in milk production is becoming better understood. We now know, for instance, that it's the lymph that provides leukocytes (white blood cells) from the mother's bone marrow to the mammary glands. Leukocytes confer 'active immunity' to the infant, enabling the baby to fight its own diseases. Further, it's been suggested that leukocytes in the milk also protect the mammary gland from infection. They're thought to do this by releasing antimicrobial chemicals and perhaps by triggering those vitally important immune cells, T cells, as well as gobbling up pathogens.[11] Think of milk as an oral vaccine for a newborn, as well as a maternal vaccine – believed to be the reason women who breastfeed are statistically less likely to get breast cancer.

A mother's lifestyle also affects her milk. For instance, compared to the average American, Old Order Mennonites, a group of farming families with a closer connection to animals and soil, have multiple differences in their breastmilk – including up to 38 different glycoproteins, and the presence of immunoglobulins that specifically relate to peanuts, ovalbumin (the protein in egg white) and dust mites (implicated in asthma).[12] Interestingly, they also have only one-eighth of the allergies of average Americans.

How much milk a mother produces is dependent on her hormones, her breasts and her baby's routine. As we learned earlier, generally a biofeedback loop means a mother produces enough for her child. Curiously enough, what she produces is also based on the gender of the child. In some mammals, the female young get a more watery, but more voluminous, milk – so, more milk, but more dilute milk. Cows, it has been shown, produce about an extra 500 litres (110 gallons) of milk per lactation for a female calf than a male calf. On dairy farms, calves are removed shortly after birth, and yet the mother continues to produce milk according to the gender of her calf, even though the calf is no longer there.

In rhesus macaque monkeys, a female baby also receives more dilute milk (lower in fat and protein), which some researchers suggest is designed to keep her closer to mum in order to drink more often, meaning she becomes more socialised.[13] The milk also has more calcium, which could help her reach sexual maturity more quickly;

an evolutionary advantage for females. The male infants get richer milk, and so can spend more time away from mum getting into mischief and being less socially aware. (If this sounds like a cliché, I don't do the science, I just report it.)

It's not just about quantity. And quality means more than just fats, proteins and sugars.

A mother's milk alters over time, tailormade to the age of her offspring. For instance, in human milk there's a sugar called myo-inositol, which is present in higher amounts just after a woman gives birth, dropping to quite low levels after a few months. Why would this be, scientists wondered – and what could myo-inositol's purpose actually be? They sourced breastmilk from China, the US and Mexico, saved in tiny amounts by mothers each time they fed their infants over the course of one year. It turned out that myo-inositol peaked when synapses were forming in the infant's brain, mostly in the cerebral cortex, the bit involved with cognition and acquiring knowledge. Synapses are the way brain cells communicate with each other, and are intimately tied to our senses – sight and hearing in particular. Scientists found that myo-inositol directly influences the formation of synapses, both in number and in size,[14] promoting the development of the infant brain.

Milk, as we've already discussed, is also swimming with living things – beneficial bacteria that confer immunity and inoculate our gut. How gut bacteria actually get into milk remains a bit of a mystery, though. We don't think they come through the mother's blood, because gut bacteria in the bloodstream can have catastrophic consequences. Take *E. coli*, for instance, which is great in our large intestine, but can cause sepsis if it gets into our bloodstream, where it can reach not only the mammary glands, but also our brain and every other organ in the body. For similar reasons, we don't think the lymph system moves these beneficial bacteria from the gut to the milk ducts, either.

In reality, there is still so much we don't know about milk, but we are now truly starting to see that its many intricate components are all meant to be in there, and they all have a role in looking after the baby – as well as the mother.[15]

So, what *do* we know? How does a mother's body know what to put into milk, and how does it actually put those things in?

Every mammal has, hardwired into their DNA, the ability to make the milk that is suited to its species. The lactocytes of an opossum know just what kind of milk a baby opossum needs. The mammary glands of a bat will extract from the almost endless list of ingredients circulating through her blood the bits she needs to turn into milk to feed her baby, along with immune-regulating compounds from her lymph. And if her blood doesn't have what is needed to make milk, she may change her diet accordingly.

In that case, does it matter what we as parents eat? Well, yes it does, for all kinds of reasons. The baby needs a diverse range of elements from milk, and so the mother needs to have a diverse diet. It's important to eat enough, and to eat well. If you're depleted in vitamin D, or K, then so is your child.

To make the exact recipe for her baby's nourishment, a mother's body will capture what it needs from food, before feeding the mother, even cannibalising the mother's body if need be. For instance, if the mother doesn't consume enough calcium or protein, her body will draw calcium from her bones, protein from her muscles, and liquidise her fat. The fat in milk – and fat is *vital* in early infancy – is also influenced by the amount of protein a mother eats. One study showed that if a human mother doesn't eat enough protein, the fat content in breastmilk fell by 25%.[16]

If the basics are right, your body can do amazing things. To a large extent it can pull apart fats, proteins and sugars and re-synthesise them. A Tasmanian devil gets plenty of protein in its diet as a carnivore, so the sugars in its milk have to be made from that. A cow, on the other hand, has to turn the cellulose in grass into high-quality protein and fat in her milk. Most of the protein a cow gets isn't from the grass she eats, however – it's from the protein-dense microbes in her stomach that help her digest the grass, and which she herself digests when the microbes die. This is how her system works.

A diet can also add unwanted things to milk, such as alcohol or drugs, which can be passed through into breastmilk, and hence the baby. Diet can also change, or add to, the protein content of the mother's milk. For instance, cow's milk caseins can enter breastmilk if a human mother drinks cow's milk – meaning a human baby who is sensitive to certain caseins can have a reaction to mum's breastmilk.

Micro-fractions of other foods a mother eats can enter breastmilk, too. For example, gliadins, found in the wheat protein gluten, can pass from the mother's gut to her blood to her baby through her milk. At this stage, the evidence shows that the gliadin amounts are too tiny to cause any irritation or a gluten intolerance in the baby. In fact, the addition of gluten to the diet of a breastfed infant after the age of six months can actually help prevent coeliac disease.[17]

What a mother eats – if she eats enough – doesn't usually change the macronutrient profile of her milk (the total amount of protein, fats and sugars), but it does change milk composition in other ways, in particular its fatty acid makeup. This has ramifications for us, too, if we drink the milk of other animals as adults. We'll come back to these in the chapter You Are What Your Cow Ate.

While all female mammals have mammary glands, you don't actually need to be born female to lactate. The thing is, just about all mammals have teats, including the males (with a few notable exceptions, including mice and horses). From an account of a lactating male goat written by Aristotle over 2000 years ago, to prison camp survivors lactating after World War II, male lactation – at least some production of milk – has been seen many times. It's thought that historical accounts of men breastfeeding could be due to a hormonal change after damage to the pituitary gland.[18] The starvation endured by prisoners, for instance, caused their pituitary, testes and other glands to shrivel. When they were rescued, and fed, the glands burst back into life, and produced lots of oestrogen and androgen, before their liver recovered enough to take those hormones out of circulation.

Male lactation is very, very rare, and usually only seen after medication, starvation, or as a side effect of cancer or obesity. In the wild, the South-East Asian dayak fruit bat, and a couple of other closely related bats, are the only animals in which we've observed fathers lactating. Some suggest they have this ability to produce milk spontaneously because they eat foods that contain phytoestrogens, natural plant chemicals that mimic the oestrogen a female mammal produces lots of. (In the human diet, soy foods are a common source of these phytoestrogens). While these bats do produce a milky discharge, others dispute whether they actually suckle their young.

Certainly, as we've seen with adoptive lactation, where a woman who has never given birth can train her body to lactate, men's bodies *can* be trained to produce tiny fractions of milk. Stimulating the nipple releases the hormone prolactin, but other hormones are required to make full milk production a reality.

Non-binary people can also lactate. A trans woman, someone who was assigned male at birth but identifies as female, can also breastfeed. While research from 2023 has shown that the milk they produce (through hormone replacement) has the exact same macronutrients as milk from those assigned female at birth,[19] whether it contains all the other micronutrients remains to be seen. Biologically, there seems little reason to doubt it's possible.

We now know that milk is more than just a mix of macro and micronutrients, more than just part of a three-way conversation between mother, child and mammary gland, while also being deeply involved in so many complex bodily processes, including our immune system. But, for many reasons, people in some countries are choosing not to drink milk as adults. This mammalian superfood is starting to be replaced by plants – for better or for worse.

It's Not
What You Know,
It's What You Drink

In 2019, a new 'ice cream' hit the market in South Africa. Though, calling it ice 'cream' might be a stretch. Produced by start-up company Gourmet Grubb, the base ingredient was EntoMilk – a dairy alternative made by puréeing soldier fly larvae (essentially maggots), which, according to the makers, 'forms a creamy liquid which looks and acts just like dairy'. The ice cream didn't taste too bad, according to reports from the time, though that could've been as much to do with the way the ice cream was made as the way the 'milk' is made. As any reader of the ingredient labels of commercial ice cream would know, dairy is just one of the many things that food technologists have put into this once noble frozen dessert – but strangely, there's been no word from Gourmet Grubb, or EntoMilk, in the past few years. Click on their websites these days and you end up getting through to an online casino. Perhaps the businesses succumbed to the so-called 'yuck factor'. Or perhaps they were just ahead of their time.

Insect 'milk' (puréed larvae) is in fact a cracking source of protein, along with some fats, iron, calcium and the like. While insect milk is still in its infancy, insect meat as a traditional meat alternative is making better headway thanks to its high protein content and lower inputs. It's said crickets can produce 11 times more human food than the equivalent amount fed to cattle. And mealworms, a type of beetle larvae, only produce one-fourteenth of the amount of greenhouse gases as beef.[1] Both are finding their way into human food chains.

Whatever your feelings about insect milk, avoiding dairy is becoming more common in developed nations. It could be for animal welfare reasons, environmental reasons, or more people identifying as lactose intolerant, or trying to cure various ills.

Like so many things – the cars we drive, the holidays we take, the clothes we wear – our choice of milk is no longer just a choice of milk. What we put in our coffee or pour on our cereal is a big part of how some of us self-identify. Our dairy or non-dairy culture is more varied, more fragmented, and based less on formal religion than in the past (though it's still often based on belief systems).

I've been talking to a lot of people about milk lately. And despite its complexity, the conversation is usually quite simplified. Often, people talk in sound bites:

'Cows are bad for the environment.'

'Oat milk has the lowest greenhouse gas emissions.'

'Almonds use crazy amounts of water and kill millions of bees.'

And they have questions:

'Doesn't soy milk contain oestrogen, so it'll make me less of a man?'

'What's the most nutrient-dense plant milk?'

'How do I choose the best milk for me?'

You'd think fake milks had only just been invented. But they've been around a very long time. Say hello to almond milk, which – to the dairy industry's chagrin – has been around in the UK for nearly 650 years.[2] As we saw in the chapter How Real Milk is Made, milk was historically viewed as 'white blood' for its life-giving properties. Due to its nutrient density, it was even considered 'meat' by some Christian faiths – and so, being 'meat', dairy wasn't allowed to be consumed on all the holy days, which made up over half the days in the year (even today for some orthodox Christians). In an attempt to keep dishes palatable, a milk substitute was invented: almond milk. While almond milk has been around since at least the 1200s, it wasn't until 2013 that its sales surpassed soy milk in the US.

And soy milk is no recent innovation, either. Doujiang, as soy milk is called in China, first appeared in texts in the 1300s, and over a century ago Henry Ford was a fan. Coconut milk has probably been around as long as humans have been growing coconuts, even if its use was often quite different from dairy. It's only lately, however, that sales of all non-dairy milks have really had much impact in Western nations. Data from 2018 suggest sales of dairy alternatives more than doubled in just six years between 2009 and 2015.[3]

It sounds like a lot, but it's still from a small base: non-dairy 'dairy' was worth about US$21 billion in 2015, compared to real dairy's US$454 billion. That's less than 5% of the market. Dairy milk sold more in a week than oat milk did in a year. But in the US in 2021–2022, real dairy rose 8% in sales, while fake dairy dropped 4%.[4]

Plant milks, in most markets, are still more expensive than real milk. That's nothing new, either. Almonds, which don't grow in the UK or much of Europe because it's too cold, were imported (mostly from what was once Persia) from the 13th century and were the preserve of the very rich. It takes quite a few almonds to make nice almond milk, and that would have been especially true in pre-industrial times. King Edward I apparently used over 18,000 kg (40,000 lb) of almonds in two years.[5] Medieval almond milk was probably thicker than today's watery concoction. According to the delightfully named Dr Neil Buttery, who writes the food blog *British Food: A History*, almonds were pounded with a touch of spring water, and often with orange blossom water or rosewater to stop the oils coming out. The paste was then steeped in more spring water, or barley water (the water taken from simmering barley), then strained. The resultant 'milk' was somewhat creamier than the stuff you'll find today. Sometimes it was curdled with vinegar to make a kind of almond curd.

I was recently in a café in Lewes, a cute castle-topped town in East Sussex, south of London. A tall young man with plenty of visible tattoos and piercings was making coffee. When I ordered a flat white, he asked, 'What kind of milk would you like? Oat … ' And then, he paused. Stopped, in fact. It seemed as if there was no other option than oat milk in my morning coffee.

I struggled with milk on that trip. As someone who milks a full-cream dairy cow, who is able to tell the season, the cow's health, what she's eaten, and which cow it is through the flavour of her milk, bought milk is not the same thing. It's thin. It has sour notes, and eggy characters. It's strangely white. But even bought milk is not oat milk. And while oat milk, when I've tried it, didn't coat my tongue like packaged soy milk, or taste weirdly out of whack with coffee, like almond milk, it's not really milk.

So, I suggested timidly, 'Cow's milk?' The man noted my age (late fifties), nodded resignedly, and picked up a lesser-used frothing jug.

Plant milks have taken off while I've been busy farming. Certainly, they seem massive in the UK, where veganism and a distrust of dairy has found more of a following than in my home country of Australia. So far. But changes are coming.

Having a rare coffee out recently in Melbourne, the bored-looking woman serving me asked what milk I'd like. Cow's milk, I explained, almost apologetically. 'That won't keep me busy for long,' she shrugged, picking up the milk carton. 'It's not even a challenge.'

When I quizzed her about it, she said the other milks are trickier to work with. They don't froth as well. Some curdle if heated even slightly too much. Many don't create that gentle foam that is the hallmark of the Australian flat white, a challenge for the barista, sure.

But this textural difficulty, and the flavour difference, is even more of a headache for the people who make non-dairy milk.

Walking up the supermarket dairy aisle, I was stumped. There were so many milks. Skim. Semi skim. Lactose free. Added cream. Unhomogenised. Organic. Now, of course, there are milks made from things you probably hadn't even heard of two decades ago, like quinoa, or hemp. (Well, we didn't call it hemp in my youth, even when we dried it in the microwave in Mrs Newman's home economics class.) The traditional dairy market is now flooded with options.

For many born prior to 1990, the first milk substitute was probably non-dairy creamer. Processed into dried form so it didn't need a fridge, these creamers were sometimes known as tea-whitener or similar. They were designed for those who were lactose intolerant. Produced using vegetable fats and later gum, some even contained a bit of dairy at one time, but usually just some of the protein, not the lactose. These creamers did add a paler colour to coffee or tea, but they tasted awful. (Some suggested they did make bad coffee taste less awful, so I guess there's that.) They were a stalwart with the International Roast coffee, Tetley teabags and Sweet'n-Low in cheap hotel and motel rooms everywhere, until the advent of tiny little plastic tubs of long-life milk.

I think my favourite thing about non-dairy creamers that I discovered while researching this book is that they can be turned into flame throwers and bombs. Like many powders, non-dairy creamer has flammable properties, particularly the sugars and fats and gums. So much so that the internet is full of videos of stupid people doing spectacular mushroom clouds and dramatic explosions using non-dairy creamer cannons. They look really dangerous. The *MythBusters* television crew only narrowly escaped injury when they tested it out. The tiny sachet you'd put in your coffee isn't going to kill you, however, not only because of the amount, but also because it needs to be tossed into the air and have a flame near it. It won't kill you if you use it as intended. At least not in the short term.

So, what is actually in plant-based milk? Some, like the almond milk by the US brand Elmhurst, have just two ingredients: almonds and water. And while almonds are healthy, with vitamin E and some calcium and antioxidant components, they're not as complex nutritionally as dairy milk. Other brands add things that milk might provide, such as more calcium, fats or protein – or things to make them palatable, or make them perform like milk. 'Barista' milks are a case in point. The term *barista*, which in Italy means a 'bar worker', is specifically used in English-speaking countries to mean a coffee geek who makes your morning caffeine hit. Now, if you simply take oats and steep them in water, or break them down enzymatically, to produce a 'milk', it probably won't froth in the same way that we're used to with genuine milk – hence the need for 'barista' plant blends.

Some 'milks' don't focus on a single flavour, but rather concentrate on trying to mimic the real thing's texture. For instance, the ingredients of Rebel Kitchen's Barista Mylk are: spring water, oats, coconut cream, sunflower oil, faba bean protein, sunflower seed extract, nutritional yeast, acidity regulator (potassium carbonate), sea salt.

Oatly, the most popular of the oat drinks in the UK and many other countries (and the inventors of oat milk, many argue), has 12 ingredients: water, oats, low erucic–acid rapeseed oil, dipotassium phosphate, calcium carbonate, tricalcium phosphate, sea salt, dicalcium phosphate, riboflavin, vitamin A, vitamin D2 and vitamin B12.

Erucic acid is considered one of the more unhealthy fats, and rapeseed (canola) is usually full of it, which could well be behind Oatly's decision to use rapeseed oil with lower amounts of it.[6] Dipotassium phosphate is a stabiliser, emulsifier and acidity regulator, which probably helps stop your oat milk from separating out into sludge in the bottom, water in the middle and oil at the top. Dicalcium phosphate and tricalcium phosphate are dietary additives, probably there because oat milk is otherwise pretty devoid of calcium, though they're also used in tablets, so they too are useful to help even out the milk's texture. The vitamins are there because without additives, oat milk is lacking much except sugar. You can tell from the ingredient list that this is the result of a food technologist, not mother nature.

MilkLab, an Australian company that makes milk specifically for the café market, has an oat milk with only seven ingredients. Their macadamia milk contains a few more things than that: Australian water, macadamias (2.5%), sugar, vegetable protein, maltodextrin (from corn), acidity regulators (340, 332), stabilisers (418, 415, 410), sunflower lecithin, salt.

Ingredient 340 is potassium phosphate, not dissimilar to the dipotassium phosphate we met above, and 332 is potassium citrate. Additive 418 is a gelling gum, 415 is xanthum gum (made by the bacteria that cause black rot in broccoli), probably used to give the drink texture, while 410 is a gum from the locust bean that also helps texturally.

Proprietary information, it seems, is quite big in the plant-milk world. Along with flavours and salt, MilkLab's soy milk contains: Secret MilkLab Soy Blend (4.4%).

Old-fashioned almond milk, the stuff those medieval Europeans used, had about 25% almonds. MilkLab has 3.5% almonds in its 12 ingredients. America's Califia Farms uses about 2.3% almonds. Elmhurst has about 9%, so that's probably how they get away with only two ingredients compared to Califia's 11.

Are more ingredients bad? It depends what they are. A curry paste that has lemongrass and ginger and coconut milk and fish sauce and cardamom and coriander and makrut lime? Probably good. Microfractions, tiny parts of a formerly whole food that have been extracted chemically in a lab and then put back into food? Probably not so good. That's because the more we do to food, the less good it does us.

When it comes to food, we all know fresh is good. Minimal processing is good. Processed is okay. And ultra-processed is a problem. Nobody, anywhere, ever, thought we should exist on a diet of ultra-processed food. The less of it in your diet, the better.

In food terms, 'processed' means it's had something done to it. And that something could often be done in a home kitchen: heating, chopping, puréeing, adding salt or acid. Ultra-processed is a different story altogether. Ultra-processed means it's had a factory-style thing done to it – chemical extractions, altering the food's basic makeup, additives that you don't have in a home kitchen, that kind of thing.

And, perhaps surprisingly considering its wholesome image, a lot of what has happened to your plant milk can only happen in a factory. Most of it, actually. If you've tried to make almond, soy or oat milk at home (and I have), they're all edible, but they don't resemble real milk in any aspect except colour.

We're not designed to digest ultra-processed food, it's as simple as that. The complex interplay of micronutrients and bioactive chemicals in any kind of fresh food – be it a carrot or a cauliflower – are usually in a form that works for our digestion. Plant-based milks, the vast majority, are anything but natural, despite some clever marketing to the opposite. They're the result of lab technicians, food technologists and factories.

From the hype, and the association the manufacturers like to make between the grains or nuts that you know, and the 'milks' they sell, it would seem that those plant milks are doing you good.

But how do plant milks really stack up nutritionally? Admittedly, nutrition is not always the driving factor for choosing a plant milk, and I'll look at environmental considerations later. But if you're taking something wholesome out of your diet and replacing it with something else, then it's good to know what it contains.

The chart below outlines the nutritional contents of some plant milks versus real milk.

Milk nutritional comparison table[7]
Grams per 100 ml (3½ fl oz)

	Cow	Goat	Sheep	Buffalo	Camel	Oat	Almond	Soy	Rice
Protein	3.0	3.6	6.0	4.5	4.0	0.46	1.0	3.8	0.17
Sugars	5.0	4.3	5.0	5.3	4.4	3.7	1.3	1.5	4.8
Fats	3.2	4.5	7.0	9.8	3.5	1.4	2.6	2.1	1.3

As you can see, real milk is usually far higher in protein, sometimes higher in fats (which may, or may not, be a good thing), and has a moderate amount of sugar in the form of lactose. The only plant milk to come close to the nutrient density of real milk is soy milk.

Let's focus for a moment on the sugar component. Added sugar, and most plant sugars, are not so good for your pearly whites. Lactose, milk's major sugar, is the least cariogenic of the fermentable sugars – in other words, the least able to create cavities. (Fermentable sugars are ones we can digest, compared to things like cellulose, which is the world's most abundant sugar found in grass and other green plants, but which we sadly cannot digest.) With its calcium and phosphate, along with some of its proteins, which also help to protect the enamel, real milk is a better option for your teeth. Taken together, milk's nutrients can remineralise teeth, while most plant-based drinks, particularly oat milk, have maltose, which is more fermentable and worse for teeth than glucose. Those sugars in plant-based milk are also bad for blood glucose levels.

The nutrient density of plant milks occasionally makes the news. A 2022 story in *The Sydney Morning Herald* quotes a dietician, Susie Burrell, who rates the nutrient density of eight milks as follows:

Full-cream dairy milk	Rating: 6/10
Low-fat dairy milk	Rating: 10/10
Soy milk	Rating: 8/10
Oat milk	Rating: 5/10
Almond milk	Rating: 3/10
Macadamia milk	Rating: 3/10
Rice milk	Rating: 2/10
Coconut milk	Rating: 0/10

The reason the full-fat milk option rates lower than soy and low-fat milk is due to dairy fat, and I think Burrell gets it wrong on that. I'm going to delve deep into fats later, but Burrell's pretty much nailed it as far as nutrient density is concerned, certainly in the simplistic sense of vitamins, minerals and macronutrients. Plant-based beverages are generally high in sugars, low in protein, and lacking in complex nutrition, even when they add some of the nutrition back in. Don't get me wrong: I'm a fan of coconut milk as it comes from a coconut. I'm a fan of it in my laksa, not in my coffee, but is it really doing me any good on its own? Probably not like milk.

I'm not alone in thinking this. In the UK, hospitalisations for vitamin deficiencies nearly tripled in the ten years to 2022. Keith Godfrey, professor of epidemiology and human development at Southampton University, believes increased plant milk consumption in the place of real dairy is playing a major part in that. Pregnant women and those of child-bearing age seem to be worst affected.[8]

It isn't just vitamins that fake milks lack. It's minerals, too – and even if they do have them, are they in digestible form? The calcium in plant milk isn't easily used by our bodies. A study in *Food Research International* from early 2024 showed that despite almonds having three times as much calcium as milk, this plant calcium is not in a very bioavailable form, so you'd need to drink 2.7 times as much almond milk compared to skimmed milk to get the same calcium hit. To obtain the same amount of calcium as a single glass of milk, you'd need to drink 7.3 glasses of oat milk, or 5.8 glasses of rice milk, or a massive 22.3 glasses of soy milk.[9]

You can like the taste of your favourite plant milk, in the same way some might enjoy the taste of Sprite. But across the board, they're not really doing you much good nutritionally. Plant milks, I mean. And Sprite.

Of course, plant-based foods do other stuff besides providing macro-nutrients. Almonds have factors in them that help prevent cancer. Oats have fibre associated with lower blood serum cholesterol. Quinoa contains 'complete' protein, meaning it has all the amino acids that we need in our diet, unlike most plant foods. But all of them, when made into long-life 'milk', aren't the same as eating the raw or minimally processed version – a handful of nuts, a bowl of porridge, or a plate of steamed quinoa.

Plant milks somehow manage to wear the halo of something natural while actually being ultra-processed, which means they are arguably at the more unholy end of what we should be eating. Dairy organisations point to the fact that plant-based drinks always want to use the term 'milk' and piggyback on dairy's good marketing.

As the authors of the 2020 report, 'Palatable Disruption: The Politics of Plant Milk', stated in the journal *Agriculture and Human Values*:

> Many mylks are marketed with more diffuse notions of health, likely because mylks lack the quantity and variety of nutrients found in dairy milk. These abstract health claims are captured in words like 'wellness' and 'cleanliness', as well as by Instagram-friendly iconography. Such promises of holistic health are epitomized by slogans like 'eat right, stay brilliant' (Rude Health) and 'feel good food' (Happy Planet Oatmilk).

As a signifier of wellness, mylks frequently gesture to what is absent. Often this is dairy.[10]

From my perspective, it seems odd that plant-based beverages, with their gums and fats and minerals added, are considered nutritionally better by many in the general public. For while nutritional science is still in its relative infancy, pretty much all sensible dietary advice can usually be boiled down to one simple constant: 'eat a variety of things, and eat the thing that is less highly processed'. The food that has had less done to it, be it wheat, chickpeas, oats or milk.

Are people trying to replace the nutrition of milk with plant-based substitutes? Yes, to some extent, but the big drop in milk consumption is nothing to do with the emergence of oat, or hemp, or potato milk.

Recent trends tell an interesting story. In 2017, roughly 92% of American households bought fresh dairy milk. And while the dairy industry in the US fell by 12% between 2014 and 2017, only one-fifth of that drop converted into plant milks.[11] In the US, liquid milk had its biggest drop of any recent decade between 2010 and 2020.[12] A large proportion of that is reduced milk drinking by teenagers, possibly because they are the most diverse generation the US has ever seen – lots of non-Europeans, and so probably also quite lactose intolerant. Consumption of milk as a beverage – the clichéd 'glass of milk with cookies after school', for example – has been dropping every decade for 70 years. As older people who drank lots of milk all their lives die, younger people are drinking less and this compounds over the years. Plant milks contribute to the drop, but they're not the main driver, analysts think.

In the UK, liquid milk sales have dropped by a third since 2000, though 96% of adults still purchased milk at some point each year. The drop is attributed to the death of the cup of tea with milk, and more recently, plant milks *are* playing a part – but curiously, only 15% of Brits surveyed actually thought they were consciously reducing milk consumption, despite a lot of households actually buying it less.[13]

Australian milk consumption remains pretty high by world standards, though the amount we drink did drop 1.1% in the last recorded year prior to going to print, 2022. We're now drinking about 6.5 litres (1½ gallons) of milk substitutes a year, compared to 93 litres (20 gallons) of real dairy.[14] At a recent count, 98% of Australian households regularly bought real milk, and a not insubstantial 42% bought plant milks. Interestingly, only about 2% of houses exclusively bought plant milks.[15]

Non-dairy milks are very keen on being seen as nutritious, and even pay for fridge space when none is needed for many of their shelf-stable products.[16] This has helped 'normalise' the products by associating plant milks with freshness and vitality, and has greatly boosted sales.

Research reported by the dairy industry shows that over a third of Australians believe plant dairy is just as nutritious as real dairy – and about one in five people think plant dairy substitutes actually contain real dairy milk. Plant-based beverages *do* trade on milk's good name, it seems.

Milk, dairy milk, can be as natural as what comes from an udder. Or, it too can be pushed and pulled, stretched beyond recognition, as we'll see later – yet even at its worst, it's not ultra-processed. However, not all non-animal milks are the result of grinding up plants. Some are made in a place that few can visit and very few can own. It's lab milk that ditches nature altogether in its attempt to replicate what walks in a field. For some, this promises to be the future of dairy. 'Real' milk, without the cow.

But before we get to that, let's take a look at where real dairy comes from, and why its reputation is far from squeaky clean.

Cows With Guns

Joey Malone was worried about my boots. We'd just crossed a couple of mucky laneways that were trafficked by dairy cows, so we could dig into the soil on Arbigland Farm – a dairy in southern Scotland he's share-farmed since 2020. Joey's job is milking cows, which means feeding cows. So, he's very keen to get good soil to grow good pasture to get good milk.

I was in Scotland, washing cow poo off my boots, because I wanted to look at Joey's three-year-old dairy. Every day, twice a day, Joey milks about 650 cows. They look like Friesians, the common black-and-white blotched breed, but these are crossed with Jerseys, which adds a caramel hint to their coats, and more butterfat to their milk. Forty girls come in at a time to each side of what is called a 'herringbone'-shaped dairy. There's a pit in the middle where the dairy hands stand, cow bottoms and udders facing them. They fit rubber-lined cups to each cow's four teats, which are linked to a pump, a chiller that cools the milk in seconds, and a big refrigerated tank.

The girls let their milk down while snacking on pellets. While those 40 cows are milked, another 40 girls come in the other side and are settled into place before the cups are swung over and fitted to

117

their teats. The average age of those cows is five or six years (though some will be 12 when the dairy is that age, Joey believes).

It takes a good two and a half hours to milk them. Each 550 kg (1200 lb) girl yields an average of 26 litres (6 gallons) per day, about 5500 litres (1210 gallons) over a whole lactation, which is about 10 months out of each year. Bigger British dairies, where the cows don't get to go outdoors, can yield 40 litres (9 gallons) per cow per day. While the dairy Joey runs was only built a couple of years ago, the herringbone technology is considered old hat. There are dairies now where the cups are fitted by computer, where the cow decides when she'd like to be milked. Dairies where her electronic ear tag is read to record her milk and what she should be fed.

Not so at Arbigland. The technology is not cutting edge, but it doesn't need to be. This is a classic dairy, like the image you may have in your head. Joey thinks very little milk you buy fresh in the UK is from grass-reared animals. His milk, often higher in fat and protein than more intensive dairies, usually ends up in cheese and butter.

I've been on a tour of the farm. I've seen the shed where the dried-off cows are put in the depths of winter for six weeks. I've seen the lush green grass the animals are kept on for the 10 months they are in milk. And I've stood in the paddock with the cows, astonished at the calm nature of a milking herd.

Each day Joey goes out into a new, uneaten paddock, clips and measures grass and dries it out in a microwave. His cows get 16 kg (35 lb) dry-weight of grass a day (plus 1 kg/2¼ lb in pellets), so accurate measuring tells him just how much of the paddock the cows need to fill their bellies, which helps him look after the soil. It might be a lower-tech milking shed, but that doesn't mean the farm isn't well thought through.

It seems hard to reconcile, from the bucolic scene I witnessed, watching those girls ruminate in the pastures, that cows are killing the planet.

Joey's farm is becoming less and less the norm. The world's biggest dairy farm is in China. At Mudanjiang City Mega Farm they milk 100,000 cows, producing 800 million litres (176 million gallons) of milk a year.[1]

Mudanjiang, which is 22,500,000 acres (a bit over 9 million hectares) in size, is about the same size as Portugal. But on this property, you'd won't see a cow from an aerial drone. That's because all the cows are inside. The grain and hay and silage that they eat are produced on 100,000 acres (40,470 hectares) of land, all grown, harvested and trucked in using fossil fuels. It's partly grown in China, but most comes from Russia. And this milk is headed back to Russia, too.

Mudanjiang has about 50 times more cows than the biggest dairy in the UK, and every cow is milked three times a day. There are 26 sheds on the farm, and the average lactating cow is just 2–3 years old when they're culled (a cow can live for up to 20 years). There are 3800 animals in just one 300 metre (1000 foot) long shed. Each cow produces 50–60 litres (11–13 gallons) a day at its peak.[2] Many of the animals come from overseas, including Australia.

Globally, intensive milking sheds like this are on the rise, thanks to increasing population *and* increased demand. Even though the Chinese are generally lactose intolerant, local dairy consumption has risen from 8.8 kg (19½ lb) per person to 20.3 kg (44¾ lb) in just 12 years. Between 1949 and 2018, annual production increased a whopping 145-fold.

This trend towards dairy industrialisation didn't start in China, and it's more widespread than you might think. When a fire broke out in early 2023 in a dairy shed about 70 miles from Amarillo, Texas – a state once famed for its grazing animals – 18,000 cows died.[3] That doesn't happen when cows are on grass. In Australia, the largest single-site dairy farm is probably Moxey,[4] which has about 1000 cows in each barn, and milks about 3700 head a day, with a capacity to milk 7000 animals.[5] And those cows aren't roaming around eating pasture, either. As Moxey – which is one of four intensive dairy farms run by Australian Fresh Milk Holdings, and is part owned by Chinese mega agricultural company New Hope Group – states on its website, 'Our cows are free to socialise, eat and drink as they choose or rest in a bed of clean fresh sand.'[6]

Until recently, the biggest dairy farm was in a country famously devoid of rolling green plains and flowery meadows, Saudi Arabia, with 104,500 milking cows at one point. Almarai started in the desert in the 1970s with 30 cows, and is now a massive industrial farming system based on the 'California Model', with shedded animals and every single speck of the feed imported. Almarai, which now has a partnership with PepsiCo, is actually multiple farms. One the *Irish Farmers Journal* visited had 46,000 head of milking cows, producing 960,000 litres (211,000 gallons) of milk every day.[7] That's a tad over 350 million litres (77 million gallons) a year. A milk tanker leaves the farm every 45 minutes. To feed these cows, both China and Saudi Arabia import feed, often from Australia and the US. Arizona, for instance, exports 208,000 tonnes of alfalfa hay every year, which consumes the amount of water that could be used to support over a million people.[8]

Dairy, I'm finding out, isn't dairy. Milk produced for human consumption can come from a house-cow such as my Myrtle, who spends her days with her calf in one paddock, and her nights in a lush green paddock all to herself. Or it may look like the dairy equivalent of a feedlot, where the cow never gets to wander in the fresh air, and the feed is all sown, grown, sprayed, fertilised, harvested and moved using fossil fuels.

Intensification isn't necessarily good for the animals. In Europe, up to 72% of dairy cows go lame each year.[9] Is it because the cows are now 700 kg (1550 lb) in size, and are expected to produce 50 litres (11 gallons) of milk a day, compared to just 20 litres (4 gallons), which was the average in 1967?[10] Are cows really just considered a fermenter on legs? Is dairy the new factory farming?

A cow, of course, can produce milk quite naturally, using only sunlight as the energy source. She will produce milk, as any mammal does, when she gives birth. Because we've bred cows to produce multiple times more milk then their (usually) single calf can drink, it's important that we milk them so they avoid complications such as mastitis. Cows like to forage for green leafy matter. They enjoy socialising and freedom of movement. Some dairies provide that. Some don't. In fact, if you manipulate the diet to be high in energy (i.e. more grain), don't allow the cows to walk much, and protect them from the normal vagaries of the weather, they produce more milk. It's more efficient, some would argue. Others might consider it cruel.

That efficiency is driving carbon emissions down, in the US at least. In 2014, 1 kg of Californian milk emitted about 1.14 kg of carbon dioxide equivalents, compared with 2.11 kg in 1964. That's a 45% reduction.[11] The US now produces 60% more milk from 30% fewer cows than it did in 1967. So, depending on your measure, intensive dairying could be seen as a plus.

The economics work against using sunlight as the main energy source. In the natural cycle of things, cows would give birth in spring, when the grass – and hence milk – is bountiful. Nature gifts us more milk in spring, when we don't fight her. But human milk demand doesn't increase in spring. As our Scottish dairy farmer Joey Malone points out, the big multinational dairy company that buys his milk, Arla, actually pays 30% less for his milk in peak grass-growing season – because rather than following the seasons, it's easier for the processing company if dairy farmers stick to grain-feeding their cows in sheds, to make the milk supply more constant.

Cows in sheds aren't reliant on grass, and for that the farmer is rewarded by being paid more for milk produced out of season. When we don't pay farmers properly, something gets hurt: the farmer, the farmed, or the farm. People, livestock, or the environment. Or all three.

Dairy's dirty secret

What happens to calves in the dairy industry is a very real problem for milk production. Every year, in order to keep producing milk, a commercial dairy animal needs to give birth – meaning something has to happen to all those babies.

In the past, the girl calves were usually kept as future milkers. And the boys? Well, if they were full-blood dairy with small muscles (less meat) compared to beef breeds (and meatier goat and sheep breeds), they were slaughtered. Killed by a blow to the head, or a bullet, or at the abattoir, on day one, or day four, or at a week old. In Australia at least, nobody could justify the cost of trucking them, bottle feeding them, fattening them. In Europe, where veal is more of a thing, some younger calves are fed out for a few months before being valued as food.

These days, thanks to AI, the old-fashioned kind – artificial insemination – the calf problem is now more manageable. Farmers can put beef genetics into a dairy cow, meaning they can raise the calves of either gender for beef. Or they can use what is known as 'sexed semen'. Using a centrifuge, the male and female sperm can be separated fairly successfully, so that instead of getting a 50/50 boy/ girl split of newborn dairy calves, it's more likely to be 90% girls.

Most dairies, like Joey's farm, take the babies away a day or two after being born (after first ensuring they get their vital colostrum) and bottle feed them. In temperate climes, the calves are kept in sheds to keep them warm and dry, until – in some (arguably more humane) farming systems – they're strong enough to end up outside on grass.

Remarkably, the stuff in the calves' bottles isn't from the dairy herd being milked in the neighbouring shed. It's usually cheaper to feed a dairy calf powdered milk than milk from its mother. From an immunity perspective, this is less than ideal. It also must be incredibly fossil fuel–dependent to milk a cow, truck its milk, process it, dehydrate it, truck the powder back in plastic-lined bags, to then have the farmer rehydrate it with warm water and feed it to a calf. Compare this to the greatly reduced carbon dioxide emissions if you simply take the milk from the cow and feed it to the calf next door, and you can see the dairy industry is quite mad in some ways.

Removing a young calf from its mum can seem cruel. It probably doesn't hurt to ask at what age a cow and her calf best cope with separation. In the process of writing this book, I separated an 11-month-old girl, Olive, from her heavily pregnant mother, Bessie, to stop her drinking Bessie's milk. Olive mooed four or five times a minute for three days, and kept mooing for a week. I prioritised the mother's health over the calf, though it didn't feel good. What does the science say, I wondered?

A recent study suggests that removing a calf at 24 hours is less stressful than doing so at 100 days.[12] Other research shows that if a calf is removed at six hours, it's less stressed than at 24 hours, which is less stressful again than if separated at four days after birth.[13] There are, of course, health and social implications for removing calves too early. But taking a calf away is usually necessary to run a dairy, and there will always be some level of separation anxiety. As Jill Griffiths points out in her book *What's for Dinner?*,[14] researchers tend to find that while a cow misses the interaction with a calf, they don't seem to miss *their* calf. Unlike a human mother who would want her *own* baby back if it was taken away, a substitute calf will do just as well for a cow. It's suggested that cows don't grieve in the same way we do when a calf disappears from their life.

How calves are separated differs with the dairy, and with the animals involved. And everything has a consequence. On our farm, our way of doing things, where we leave the calf on for half of each day, has been shown to reduce milk yield to the farmer by a whopping 42%.[15]

Hormones in milk

Some people are worried about the fact that there are hormones in milk. And it's true, there *are* naturally occurring hormones – but they're such big molecules they get broken down in the human gut and rendered inactive. We know a lot about them because in some places, the Americas mostly, they actually inject dairy cows with hormones to get them to produce more milk.

Bovine somatotropin (bST) is a naturally occurring cow growth hormone. And researchers a few decades ago worked out that if you inject a dairy cow with bST, it boosts her milk supply by 10–15%. Originally, however, it took 200 dead cows to get enough of the hormone from their pituitary glands to give to a single dairy cow. Enter science, and the boffins worked out how to synthesise a biologically identical hormone to bST, called recombinant bovine somatotropin, or rbST.

It sounds like the kind of efficiency we want. More milk, with the macronutrients pretty close to the same. Unfortunately for the cow, rbST has been found to increase the incidence of mastitis, an inflammatory disease of the mammary gland, by about 25%. It increased the risk of a cow not being able to get pregnant again by 40%. And cows showed a 55% increase in lameness.[16] No wonder rbST is banned in Europe, the UK, Australia and New Zealand.

Other research shows you can improve production in dairy cows just by caring about them enough to give them names. It's about a 5–10% increase in milk, so a more modest improvement in efficiency, but at least it doesn't send cows lame. Or you could play them music, which can apparently boost milk production by about 3%, with no noted increase in mastitis.[17]

Grazing land vs arable land

There's a few things, environmentally, that give dairy a bad name. It's part of the reason some people have started to seek out plant-based options. To certain commentators, grass-based systems that use ruminant animals (those with four stomachs, such as cows, sheep and goats) are the single most destructive things on the planet. They use lots of water. They take up lots of land. And they produce greenhouse gases from their guts and from their poo.

In terms of land use, chopping down trees to make way for dairies is a bad thing. And yes, in a Saudi Arabian summer, each lactating cow will need about 300 litres (66 gallons) of water a day. Even though 50 litres (11 gallons) of that water comes out as milk, that 300 litres of water could be used in another way.

But cows don't have to be kept in sheds. About a third of the world's usable land mass is grassland (or pasture, or range land, or whatever name you'd like to give it). It's often land that doesn't grow crops such as soybeans, peas, spuds, carrots or cabbages, but it grows grass quite successfully. We humans can't digest grass – but a cow (or other ruminant) can upcycle that grass into her milk (and into meat, too).

It's been suggested that it takes about 628 litres (138 gallons) of water to produce a single litre of milk. If this is rain, and it falls on grass, then it's not like that water is stolen from your shower. But could this water be used for other things? Like growing crops? Possibly. It depends on where that is.

In New Zealand, the use of water has been analysed on pasture-based dairy farms. Researchers found that there's water that falls as rainfall on grass (green water), water taken from rivers and bores (blue water), and a thing they call 'grey water' that is needed to water down pollutants in the system. And the result? New Zealand dairy farms have been shown to use about 680 litres (150 gallons) of green water (rain), and just under a single litre (945 ml) of blue water (irrigation water) to produce each litre of milk. The problem is the grey water. To ensure that the nitrogen run-off from a farm isn't polluting waterways, it can take over 11,000 litres (2420 gallons) of water to produce a litre of milk.[18] New Zealand dairy farming is, as a whole, reliant on artificial nitrogen fertiliser, and perhaps doesn't take enough responsibility for the huge amount of nitrogen-rich effluent produced from the dairies – the cow poo. Nitrogen pollution is one reason nearly 60% of all rivers in the country are now considered unswimmable.[19]

Like all things, too much of something – in this case, cows and nitrogen fertiliser – is damaging. But dairies can be better managed, and the amount of water controlled. As we've seen, dairy isn't just dairy. And not all dairy farms are created equal.

Land use isn't the same everywhere, either. Cow's milk is usually identified as a big user of land, but it takes anywhere between 1.18 and 54 square metres (m^2) to produce one litre of milk. The lowest value for cows is about twice the amount of land used for the highest plant-based beverages: oat milk and soy milk (both at 0.66 m^2), then almond milk (0.50 m^2) and rice milk (0.34 m^2). Of course, a cow can be fattened on land that isn't suitable for growing oats or almonds or soy. It's complicated, and not just about land area, but land quality.[20] (In Australia, for instance, only about 4.1% of the land is arable – that is, suitable for growing crops rather than grass.[21])

Global averages and big data can be – and often are – skewed to make plant milks look good and animal milks bad. But there's a big difference between farming systems. The carbon emitted to produce dairy, for example, can vary by a factor of 10 (meaning some dairy systems produce 10 times as much emissions), and land use by a factor of 50.

When we look at global numbers, it pays to be cautious. The largest increase in dairy consumption is in the Asian subcontinent, but it's also home to some of the least 'efficient' cattle. But we have to realise that not everywhere looks like Scotland or Montana, or California or even Tasmania. For instance, while in the next few years it's predicted that a third of the global dairy herd will be based in Africa, they're also only expected to account for about 5% of world milk production.[22] This isn't mega farms using the 'California Model', though. It's farmers using land and turning plants into milk. To you and me, and people in cafés in Lewes, this may seem inefficient. But those dairy animals are also vital to small farmholder's *lives*, not just livelihoods. They're upcycling nutrients and providing not just milk, but meat, to nations whose recent history is blighted by protein deficiency.

Cow burps

Unlike an almond or an oat, one thing a cow or goat or sheep *does* do is burp methane – a greenhouse gas that, depending on how you measure it, has a heating effect on the Earth between 8 and 25 times higher than carbon dioxide over 100 years. It's a short-lived gas, which lasts about 10 years in the atmosphere before being broken down by sunlight or biological means. The carbon in the methane a cow burps out comes from the food it ate – mostly grass, in the case of Joey's dairy herd.

Methane is made in the fermenting foregut of ruminants – those with a rumen, including all the dairy suspects cows, goats, sheep and buffalo (plus the three-stomached camel, a so-called 'pseudo-ruminant'). The large rumen is a wondrously biologically active place, packed full of helpful microbes that allow the animal to digest the cellulose in grass. Grazing animals such as horses and kangaroos that lack a fermenting rumen need to eat about 60% more grass to extract the same amount of energy from it – so, they use more land to produce the same amount of milk, or meat. (And before you say we don't eat horses, some people do. A not insubstantial 4.3 million of them a year, or just under 800,000 tonnes of meat. And it's also worth talking about because of how much land they use if they're not eaten, which is land taken out of agriculture.) The methane comes from certain types of fermenting organisms that are really quite useful in the biology of a cow or other ruminants: archaea, single-celled organisms that predate bacteria.

Methane is a powerful greenhouse gas, responsible for about 30% of the increase in global temperatures. About 40% of the release of anthropogenic (human-caused) methane emitted each year comes from the fossil fuel industry. (And according to the International Energy Agency, methane output from fossil fuel energy companies – mining, fracking, drilling – is also probably about 70% greater than reported by national governments.)[23] For instance, a study from early 2024 shows that methane from US natural gas drilling is nearly three times higher than official US Government estimates.[24] A bit less comes from all agriculture, with cattle and other ruminant farming (mostly beef) contributing the lion's share of that.

If the size of the global dairy herd remains unchanged, the biogenic methane (the cow burps) won't contribute to more global warming – unlike what comes out the back end of your petrol or diesel car. The bigger problem comes if we keep expanding the dairy herd. To say drinking milk has the same effect on the climate as taking a short-haul flight or driving your car is a false equivalence.

Because of both the methane it emits, and its fossil fuel use, dairy is considered a contributor to global warming. Some estimates suggest that, globally, about 3% of anthropogenic greenhouse gas emissions come from the dairy industry. According to an article in *Nature Climate Change* in 2022, however, rice will heat the planet more than dairy by 2030.[25]

How much of dairy's contribution to global warming is due to methane, and how much is due to the fossil fuels required to power the dairy, truck the milk, pasteurise the milk, package it, ship it and refrigerate it? These numbers matter, because if you replace milk with something else, that replacement may also need those fossil fuels to sow it, fertilise it, truck it and process it. In terms of packaging and transport to the end user, the costs (environmental and in carbon dioxide emissions) are pretty similar, whether it's real milk or not. I'll say that again, because you don't read that on the side of your rice milk carton: apart from the cow, the emissions are pretty much the same, if not more, for plant milks.[26] But fresh milk of any kind does take more refrigeration.

If you look only at methane, dairy cow burps account for 0.7% of total US greenhouse gas emissions,[27] and 0.29% of Australia's total.[28] In the UK, although they've reduced enteric (burped) methane (11% more milk with 24% fewer cows), they also don't dig for coal or drill for gas as much as the US and Australia do, so dairy burps are responsible for about 3% of the nation's total greenhouse gas emissions.

Is 0.29% or 3% of emissions going to make you put oat milk in your coffee, while still buying fast fashion, burning gas, travelling to

Tuscany and renovating your house? Consuming stuff, essentially. Avoiding dairy (or meat) simply isn't the best thing you can do for the planet, no matter what the internet tells you. It's *one* thing you can do, and it may make you feel powerful in the moment, but it's not the important thing. Reducing fossil fuel use is. Farmers like Joey Malone and others who grass feed cows get tired of being lectured on emissions by companies such as Oatly, whose majority shareholder is China Resources, a Chinese government owned entity that also heavily invests in coal mines.[29]

All food, all eating, has an impact. But good farming can repair that impact. What contributes about 15% of our global greenhouse gas emissions? About five times more than dairy? Food waste, that's what.[30] If you buy a takeaway coffee with plant-based milk in a compostable cup, and don't compost that cup but throw it in the bin, you could well create more methane than the cow's milk that could've been in the coffee in the first place. In fact, just making the cup (without a lid) emits about as much greenhouse gas as the pasture-based cow's milk in a flat white,[31] and the methane from throwing that cup in the bin just adds to the problem.

The Earth is designed to grow food. It's the one thing we know we can do forever, if we manage it well. We can't mine forever. We can't frack forever. We can't even build electric cars forever, because they all consume things that are finite. In human terms, what we truly need can come from soil, air and water with the addition of sunlight. If dairy comes from the sun, it is more sustainable than any plant-based milks that rely on factories to produce them. And much of what builds and powers a factory is not energy from the sun, but fossil fuels.

Of course, most modern dairy in developed nations isn't just powered by the sun. It relies on machines to extract the milk, energy for transport, and energy in the production process, including packaging and refrigeration. There's also the question of the diet of the cow, and whether that carbon coming out as methane is a result of fossil fuels, too, thanks to artificial nitrogen use, and grain and silage and other supplementary feeding.

There's a way that dairy farms can help the climate, too: by drawing carbon down into the soil. Carbon in oceans is bad. Carbon in the atmosphere is causing climate change. There's carbon in all living things, too. That carbon cycles, between the air, into plants, into us, into soil, into air again, in a loop. Farmers can, and do, store carbon in soil. In fact, soil is the second largest carbon sink after the ocean.

When I contacted Tom Gregory at his dairy farm in Somerset in the UK, he was very excited about carbon. A young dairy family, Tom and his wife Sophie and their kids have 200 head of organically run dairy cows. How do their emissions stack up? Well, being soil nerds and passionate about creating a cleaner, greener future, they're doing quite well. On each of the 3500 hectares of land they farm, they have stored 11 tonnes of carbon dioxide since they took over, by doubling the soil organic matter. This is carbon that is taken out of the atmosphere by a living plant, namely grass, and put into soil using clever farming. And cows.

Tom isn't just making stuff up. His carbon audit of the farm says he emits 1.02 kg of carbon dioxide per litre of milk he produces. The Gregory's farm emits about 5–6 tonnes of carbon dioxide per hectare, making it a net 5 tonnes of carbon dioxide they've stored in soil. Tom's farm is carbon negative, sequestering 2100 tonnes of carbon dioxide more than it emits every year. He's carbon negative, which means climate positive. Buying Tom's milk is good for the planet. And farms like his could well exist near you, too.

Bugs in soil, eating seaweed and more

Methane is one accounting tool. And emissions are one measure. But well-managed cows (and more specifically, dairy farms) can not only store carbon in soil, they can also play a role in helping to oxidise (essentially decommission) some of the methane they produce, using soil life to do so.

Methanotrophs are single-celled soil organisms that are the only known biological method to oxidise methane into less warming carbon dioxide. Soils represent about 6% of the potential methane sink, with the rest oxidising in the air using sunlight. Methanotroph research is hard to conduct, but there is evidence that grazing increases methanotrophic digestion of methane compared to ungrazed areas, particularly close to the soil's surface.[32] It also helps to only moderately graze the land, and have porous soil. Good farming, it turns out, can increase the rate at which methane is broken down by living soil. Not so much that it can break down all the methane from fossil fuel emissions, however.

There's also plenty of work being done to decrease methane emissions from cattle. Asparagopsis, a red seaweed developed in Australia's north, is believed to reduce methane by up to 90%. It reduces the activity of things called methanogens in the cow's gut, and increases the animal's ability to digest its food. This actually leads to an increase in milk production, because of a higher feed conversion. The active ingredient, bromoform, has long been known to reduce methane, though its carcinogenic properties put a hold on research using its pure form until this natural form was discovered.

In other work to reduce emissions, some cattle are fed nitrogen, which lowers the amount of methane emitted from the rumen, although it's problematic for cow health. Other companies and farmers are trialling garlic, enzymes, oregano, green tea, cinnamon and Agolin Ruminant, a proprietary feed supplement, actually an oil with a herb and spice blend added. All of these are proven to reduce enteric methane, the stuff burped out by cows.

And, of course, there's genetics. We can actually breed for cows that produce less methane. At the time of going to print, there's already semen from lower-methane bulls on the market, promising methane reductions in subsequent cattle generations of about 1.5% a year, translating to reductions of up to 30% by 2050.[33]

Dairy, like all industry, has to work out ways to reduce emissions. And there's strong motivation. New Zealand, a massive dairy exporter, is bringing in a methane tax in 2025, so farmers will have to pay per cow. Ireland, like many countries that have signed up to reducing their carbon footprint, is looking at reducing methane by culling dairy cows. It's not their petrol cars or diesel tractors that will be cut, it's cows that face the chopping block. Maybe that's a good thing for the planet. Maybe it's short-sighted.

Nothing in agriculture is simple. While killing all the cows sounds like a cracking idea to some, most people who actually work to save the planet know it's more nuanced. For instance, an article in *Nature* in 2022 showed that you can actually reduce emissions, and feed more people, using ruminants rather than single-stomached animals such as pigs and chickens.[34] The grain used to feed pigs and chickens comes at a cost. Replacing only 12% of pigs and chickens with ruminants would reduce greenhouse gas emissions by 5% and feed over 500 million more people.

We can also grow other crops in some areas that currently run dairies. But what would we grow? One dairy farmer I met who leased part of their land to grow crops found he made twice as much money off dairy as wheat or grains. When I pointed to the potatoes on his farm, he wasn't convinced they could be grown without long-term damage to his soil, even using crop rotations. What's more, his wife wouldn't eat those potatoes, which were headed to McDonald's. 'I've seen what they spray them with,' she sighed.

How you view a cow's role probably depends on whether you grow food or not. Or if you're in a rich country or not. The people most wedded to the removal of animal agriculture are usually urban, white, and wealthy. One in every four farmers in the world owns at least one dairy animal. To remove dairy from the planet would inevitably harm those most vulnerable. It might make sense if you live in London or New York or Sydney to use a plant-based milk instead of the real thing. But it may not make sense if you live closer to the place where your dairy is produced.

Truly local

Twelve minutes. That's all. Twelve minutes out of Wellington, New Zealand's capital, sits a goat's milk dairy. If you want to see local milk at local scale, then perhaps this is it.

Brooklyn Creamery is a micro-dairy, using predominantly Saanen goats that forage in the ragged peaks near the wind turbine on Wellington's outskirts. The milk is pasteurised on site, put into re-usable (and importantly, refillable) glass bottles, and delivered into town the same day. Ten per cent of the takings are put into Caprines for Conservation, a goat-funded ecological trust to help restore native habitats and bring back New Zealand's iconic bird, the kiwi, to the farm. When I visited the farm with Naomi Steenkamp, who runs the dairy with her partner, Frans, the goats were coming and going from a shed that allows them to shelter from Wellington's famously inclement weather. Naomi doesn't see the goats as doing it alone. It's a team effort: farmer, farmed, farm. All things working ecologically, and ethically, to produce food that hasn't done millions of miles or burned litres of fossil fuels. Together with their four children, the Steenkamps are trying to make local milk really local, and prove it can be done with environmental credentials. Meanwhile, the oat milk some Wellington residents drank on my visit in 2023 had all been imported.

All over the world, dairy is a thing, feeding farmers while also creating nutrient-dense food for the rest of us. Dairy doesn't have to be big. It doesn't have to be destructive. Dairy doesn't have to poison rivers, cut down forests, burn fossil fuels or ignore its role in climate change. It really isn't the goat, or the cow. It's the how.

Rebel Mylk, whose nine-ingredient Barista Mylk we met earlier, seems to be more into buying carbon offsets than reducing emissions. They say:

Offsetting our carbon emissions is so important when we're sourcing ingredients for this amazing recipe from all around

the world, making Barista Mylk in the United Kingdom, then shipping it to countries like Australia.[35]

Apparently, 'every carefully selected ingredient is minimally processed and as close to [nature] as mother nature intended'.

I'm biased in favour of dairy, I know. I'm biased in favour of cows, of that there is no doubt. But I'm also obsessed with the numbers, and with feeding the world. There are plenty of people who care about the planet who just think we need to rethink how we do all things – transport, energy *and* farming – not just discard things we know can be done forever, like grow plants and animals. We can electrify everything, and use wind and solar as our power sources. We can build our houses and flats better, to be more energy efficient. And we can use technology to our advantage. Farming is no different, and good farmers like Joey and Tom, along with tens of thousands of others around the world, are working out how to grow food and not destroy the Earth. Taking aim at dairy as a major culprit is the easy option for those who haven't really thought things through and who usually grow nothing.

There are 60 million horses in the world, an inefficient feeder using up tens of millions of hectares of land, millions of tonnes of grain, billions of litres of water, and hardly being used to feed anyone. But nobody talks about killing the horses. Hundreds of acres near me are given over to alpacas, with most of the wool being used to mulch trees, not clothe humans, and the meat rarely eaten. In the US, domestic dogs and cats drive climate change as much as a meat-eating human, each one causing twice as much greenhouse gas emissions as the family car. You can point the finger at farmers, but a single AI program (GPT-2) emits about 300 tonnes of CO_2 emissions when it is being created – the same as 125 round-trip flights between New York and Beijing.[36] Meanwhile, the most irrigated crop in Australia is the domestic lawn.

It's easy to blame agriculture for the world's ills. But we have to eat, and there are other climate culprits that few seem to talk about.

Home air-conditioners, for instance, will account for three times more global warming than dairy between now and 2050.[37] Humans survived 200,000 years without air-conditioners, but not without food. And air-conditioners can't store carbon in soil.

Dairy does have an impact. It varies enormously from place to place, from farm to farm. But it's not inherently evil, in the same way growing oats isn't inherently evil – although it must also be said that growing oats usually depletes soil fertility, burns fossil fuels, uses arable land that could be used for other crops, is often a monoculture, and just before harvesting, the oats are often desiccated with glyphosate (Roundup). (If you want to avoid glyphosate, best to get organic oats, which can have 1/40th of the chemical residue of conventional oats.) Lots of oats in the US also contain chlormequat, a chemical plant growth regulator used by grain farmers, which is also an endocrine disruptor; one that reduces fertility and harms foetuses in utero. In 2023, researchers found chloremequat in an incredibly high 90% of urine samples of Americans, way more than they had previously found.[38]

All things have consequence. Ditching dairy could well prove to be a bad idea. Certainly, if it's replaced by ultra-processed foods grown in monocultures that require laboratories and factories in an effort to mimic natural nutrition, then it could be a really dumb idea.

You can make your own mind up about what you want to eat, just base it on sound facts, or admit you just feel better one way or the other and don't care about the numbers. Or only care about the cows. It makes no difference to me if you drink milk or eat dairy or not. If you're old enough to read this, then you're adult enough to make your own food choices, and some people will have a belief system that an old white bloke like me is unlikely to change anytime soon. But take it from someone who grows fruit, vegetables and nuts, as well as milks a cow – and has read everything they could muster on the issue – that dairy animals aren't the worst thing that can happen to the land or the planet.

There is a way to make parts of milk without cows, however. For that we'll need to head out of the paddock (and out of the 3800-cow shed), put on our white coats and hair nets and head into the laboratory.

Don't Have a Cow!

Phil Newton peered out through his black-rimmed glasses, looking at me like I was an idiot. He was taking me through his pigeon shed, showing me the squab – baby pigeons – tucked into nesting boxes. 'The adults feed them milk for the first week,' he has just told me, grabbing a featherless, fat, only-a-mother-could-love butt-ugly chick and holding it up to my face.

'Milk?' I asked, gormlessly.

'The parent birds make pigeon milk, and feed it to their babies,' the pigeon breeder said. 'They spit it into the baby's mouth.'

Pigeon milk is a thing, apparently, with the chicks being fed so-called 'crop milk' before they move on to solid food. Crop milk is firmer than mammalian milk, more the texture of cottage cheese. It's pale golden in colour, and comes from the crop – an organ in the neck of the bird that grinds grain down into digestible pieces. But it's not half-digested seeds that the pigeon is spewing up. It's a really high-protein, high-fat substance made by the pigeon to nourish their young and help with the chicks' immunity.

Like a mammal's, pigeon milk contains immunoglobulins and beneficial bacteria. And its production is triggered by the same hormone, prolactin. Both pigeon parents make it, however, not just the female, and pigeon milk doesn't contain carbohydrates (things like lactose). It's not so much made *by* the cells of the crop, but by the *destruction* of crop cells, we think. The bird regurgitates this delicious phlegm into their chicks' open mouths. Baby pigeons almost always die without crop milk, which is one reason why the pigeon meat industry is very hard to intensify. Crop milk has helped pigeons avoid becoming the new chicken.[1]

Pigeons aren't the only birds to make 'milk'. Flamingos make it, too, though theirs is red in colour from the shellfish they eat. Emperor penguins also make it, though just the dads. It's a form of co-evolution, each of these birds developing the ability completely separately from each other, and separately from mammals.

Is pigeon milk actually milk? It seems it's very hard to define this simple substance. And even harder to copy it.

If plant-based 'milks' are trying to mimic real milk, they could find it hard going – partly because we don't yet really know what milk is, let alone exactly what's in it. It's hard to replicate when you don't know what you're replicating. And so far, while some plant juices might perform closer to milk in your cappuccino – far closer than pigeon milk, I imagine – those plant milks are generally bereft of real milk's macronutrients, let alone its micronutrients.

There's other work being done, scary work, to see if we can alter the milk an actual animal gives. In China, cows have had human mammary gland genetics inserted into them, so their milk is '80 percent the same as human breastmilk', according to the researcher involved. The 300 cloned cattle on the researcher's farm near Beijing, according to reports from 2011, 'could provide the same nutritional properties as human breastmilk, but with a taste even stronger and sweeter'.[2]

There's also a move to make milk that is nutrient dense, particularly higher in protein – without even using a cow, or a pigeon, but still using genetic modification.

This technology uses genetically modified yeasts, the kind being harnessed to provide proteins to put in fake meats. It's been dubbed 'precision fermentation' by proponents, who point out that it's the same basic process – fermentation – that is used to turn soybeans into soy sauce, and barley into beer. The 'precision' bit is marketing spin, and clever it is, too. It's also called 'recombinant protein expression' or 'microbial fermentation' or 'cell factories', depending on who you talk to.

In Australia, our pre-eminent scientific organisation, the CSIRO, does a lot of cutting-edge research, these days often in partnership with private companies. One such partnership is with start-up company Eden Brew. They're inserting into yeasts the genes that make cow milk proteins, to ferment a milk substitute. As they explain:

> Nature's method is perfect ... but it didn't count on having 10 billion mouths to feed. So we've stepped in. We map cow milk genes to produce dairy proteins that deliver cow milk nutrition and sensory experience.

Sounds fair enough.

> Our process of fermentation takes the gene map and uses a natural yeast to grow them into nature-identical cow's milk proteins.

Eden Brew's founder and chief executive, Jim Fader, reckons the company's business model will be based on Coca-Cola's supply chain.[3] The idea is that you make a major component in one place, in Eden's case the protein, and then ship it around the globe to water down and make it into the finished product. Apparently, Eden Brew will ferment, then dry and sell, a product that has six of milk's most abundant proteins.[4] They don't intend to end dairy, they say, just complement it. Eden has even attracted a dairy consortium, Norco, as an investor, with product expected to hit the market from 2028.

Eden Brew is hardly the first in this space. Launched in 2014, California's Perfect Day, backed by the actor Leonardo DiCaprio and Walt Disney CEO Bob Iger, is already making a protein base that is used to create animal-free whole milk. This protein base is in Nestlé's Cowabunga, released in the US in late 2022, using casein proteins from fermentation, along with oat concentrate and high-oleic sunflower oil. Perfect Day's protein powder is also in chocolate, thanks to another huge international food processor, Mars.

Another early player in the non-animal-milk world is Singapore's Very Dairy. They say:

> **All Very Dairy products are made from animal-free dairy. It's the same nutritious and delicious goodness you find in traditional cow's milk, but made without animals.**[5]

Very Dairy's genetically modified yeasts are bred up in large fermenting vats using vitamins and sugar to produce a 'milk protein that's identical to the one produced by cows'.

I'll fact-check these nutritional claims in a minute. Their wording is important.

Very Dairy's whole non-dairy milk has the macronutrients of protein, fat and sugar in proportions similar to real milk. The ingredients are as follows:

> **Water, non-animal whey protein, maltodextrin, canola oil, sucrose, coconut oil, permitted stabilizers and flavouring, and calcium potassium phosphate citrate.**

To be honest, the list does seem reminiscent of the ingredients in many plant milks – something ultra-processed. Remember, if you can't make it at home, it's generally something a sensible nutritionist wouldn't recommend you consume much of.

It's more than just milk these corporations are replacing. In Germany, a company called Formo is on a mission to replace cow burps with factories fermenting the milk components and making cheese. They say, 'this is no terrible tofu vegan cheese affair', trading off the fact that vegan cheese is, well, to a cheese lover, pretty vile to eat, even if it may boast a smaller ecological footprint. The photos of the cheeses on their website (the real things were yet to be released at the time this book went to print) look like the real deal.

The Washington Post reports that there are nearly 30 companies working on similar animal-free dairy goods around the globe, though only a small amount of final product has so far made it to market.

Brave Robot, an American company, is already using Perfect Day to make ice cream. In early 2023, a cream cheese product using Remilk's protein and made by General Mills (one of 11 food mega-companies that produce 70% of the world's ultra-processed food) was launched in Minnesota.[6] In their marketing material, they boast: 'We produce milk that is 100% identical to cow's protein.' It's not wrong, as such, but perhaps not the whole truth.

And there's plenty more on the way. Change Foods is a US/Australian tech firm looking at making cheese. Britain's Better Dairy is about to launch a non-dairy 'cheddar' and 'stilton' made using fermentation, and want to replace all dairy with lab milks, while the Kraft/Heinz–backed company New Culture is due to launch a mozzarella and other animal-free cheeses in the US at high-profile chef Nancy Silverton's pizza restaurant, Mozza.

Even traditional cheese masters are getting into it. A company called Standing Ovation is hoping to do the vegan cheese thing in France, with backing by the Bel Group, a French cheese marketer with international reach.

Some of this innovation is run through The Kitchen FoodTech Hub, part of the Strauss-Group that controls about 12% of the food market in Israel. Strauss-Group is working with global food mono-liths PepsiCo, Mondelez and Danone.[7] It shouldn't come as a surprise that these companies are also in that list of 11 that make most of the highly processed food many of us already eat.

Lots of companies are skirting on the edge of this new science. But copying natural systems – 'nature inspired technology' – is really hard. We know, for instance, that we can use a genetically modified yeast, *Yarrowia lipolytica* (don't worry, you don't need to remember the name), to produce a human milk fat substitute, a thing called a triacylglycerol.[8] But we feed the yeast palm oil and glucose, both of which have to come from somewhere else, so it's not without consequence either.

Making some singular proteins from yeasts is not that hard using modern technology. But making the complex casein molecules – micelles – the way they exist in milk, for instance, is not so easy, and nobody has done it at scale. As one scientific paper from 2022, looking at the viability of precision fermentation, says: 'the ability of making such micelles from recombinant caseins has not been shown'.[9] So, making the complex molecules you get in real milk is harder than you'd think. Can we make milk proteins cost effective and as complex in reality as the very thing they try to copy? Scientists aren't so sure. We aren't even sure of the environmental implications.

A lot of companies are big on talk, but light on responses to my queries. They also know how to market themselves, even before they have anything to bring to market. Change Foods, a US-based company that built a factory in the United Arab Emirates, for instance, reckons they can create casein with one-tenth of the water and one-fifth of the energy of conventional dairy (note, just the casein, not the cocktail of other nutrients that dairy also provides). But at the time of writing, nothing has yet been put on the market – apart from the opportunity to invest.

If I had to pick one brand that does the precision fermentation sell really well, it's probably Remilk, a company that reckons they can outcompete cows using their fermentations. 'Same dairy, minus the cow' is the mantra they use, not dissimilar to the tone that many companies echo. Remilk is 'creating dairy that is a far superior version of itself' and is 'identical to its traditional counterpart!' They also talk about attracting board members from

PepsiCo, Danone and Nestlé, so you know they care about human health and wellbeing.

Corporate morality aside, what can we make of the nutritional claims by animal-free dairy manufacturers? As I noted before, these are ultra-processed products, the thing a sensible nutritionist will tell you to avoid, or at least minimise. But what do they contain that's *good* – and what *don't* they contain that they should?

A major protein group they are producing is casein, which is usually in complex molecules called micelles in real dairy, which as we've just seen are very hard to replicate. Casein can be good, but it's also an allergen. In fact, many people who think they are lactose intolerant but haven't been tested could well have some trouble digesting casein, the major protein in cow's milk.

The extra ingredients in animal-free dairy – things like the fats they add, the emulsifiers and stabilisers – all have a role in the product's impact and nutrition. I'll look at fats in detail later, but to summarise here, the use of sunflower oil can be fraught nutritionally, while palm oil and soy oil are certainly problematic environmentally. Canola oil is probably not bad, as rapeseed (canola) has been bred to contain more of those healthy omega-3 fats – although when it's grown as a monoculture using herbicides and artificial fertiliser, one has to wonder about its impact on long-term soil fertility.

The sugars they add to lab milks are also problematic. Lactose, as we've seen, is good for dental health, but two-thirds of the world's population have trouble digesting it – so the companies making animal-free dairy tend to add maltose or dextrose or sucrose (table sugar), which are all considered inflammatory to cells and bad for teeth. Eden Brew, for instance, adds a small amount of table sugar to their lab milk to mimic the sweetness of the real thing. Bored Cow, another California-based animal-free dairy – which says it offers 'a new kind of milk alternative made with real milk protein from fermentation rather than factory farming' – uses sugar, as well as sunflower oil and added (and unnamed) flavour. Meanwhile, the addition of oats to Nestlé's Cowabunga milk increases the maltose content – one of the worst sugars for your mouth, and the best for your dentist. Maltose also spikes blood sugar, even more than glucose does, so it's not ideal for anyone with blood sugar issues, such as those who are pre-diabetic.

So, what about protein? Surely the new animal-free dairy nails that? Eden Brew is concentrating on fermenting two cow proteins at present, building up to six before going commercial. They may even get to a possible 20. Most of the other companies don't state how many proteins they produce, and the ones I contacted didn't respond to my questions. But, because it's so expensive to set up a single bio-reactor, and you need one for each protein, they'll probably produce just one or two forms of casein – the dominant and perhaps most structurally important protein that we miss in plant milks and love in real milk and cheese.

One or two proteins, then. How many proteins are in actual cow's milk, do you think? It turns out more than six. More than 20, even. The complex interplay of the mammary gland of the cow and her diet, along with her interaction with the ecosystem, means her milk can contain 4654 different proteins over her lactation, according to a 2020 report by French researchers.[10] So-called 'precision fermentation' to produce protein is far simpler than the name suggests, and will only provide about a half of one percent of the complexity of real milk. That's 0.5%.

In terms of greenhouse gases, it's a bit early to tell just how much factory-made non-dairy milk proteins will contribute to emissions, but very recent research shows its impact is pretty much the same as when we extract protein from cow's milk to use in food processing.[11] Both have the chance to reduce emissions over time, but there's no environmental benefit there.

Will this stuff taste good? Maybe, but it'll be a big ask. While science is a wonderful tool, letting scientists in the kitchen is hardly a recipe for gastronomic success. (Probably about as successful as letting the chefs run the labs …) And looking at the nutritional data being gathered by other scientists, it's hardly a recipe for truly nourishing, truly healthful food by comparison.

Don't get me wrong. The use of lab-derived milk proteins is a great boon for industrial food production. Some of the milk we currently produce is pulled and pushed and extracted for use in ultra-processed food, and getting rid of the cow to make more barely nutritious food is a noble goal, so long as it doesn't supplant unprocessed or barely processed food.

About 18% of the world's milk is used in powders such as whey powder, skim milk powder and whole milk powder, and much of that ends up in ultra-processed food. Replacing that milk using factories instead of factory farms would be a positive. Getting rid of intensive dairies where cows live for barely three years, where their food is all trucked in – so their animal welfare, environmental sustainability and fossil fuel use are all questionable – is probably a good thing.

Replacing the dairy animals that, globally, one in four farmers use to upcycle nutrients and feed themselves and the world? Well, perhaps that's a different thing.

Because if fake milk – even using cell factories – is only replacing a tiny fraction of all the many complex, health-giving elements in real dairy milk, this could have real-world ramifications that may take a generation or two of human eaters, and their doctors, to fully comprehend.

But first, let's look at some unintended consequences of replacing dairy with fake milks, and what that means for the world.

Love Oatmilk?
Good. Someone
Better Love Bacon

'It's more than four times as nutritious as cow's milk,' exclaimed the scientists. They were talking about research on a new milk source.[1] Cockroaches, it turns out, secrete a milk-like substance to feed their young, according to the story 'Scientists Think Cockroach Milk Could be the Next Superfood', which appeared a few years back on the online platform Science Alert.[2]

If your definition of milk is something like 'a nutrient-dense substance secreted by the parent to feed its newborn', cockroach milk fits the bill. The milk crystals that the cockroaches produce are really protein dense, and they also contain fats and sugars, just like dairy. Scientists are trying to work out how to harness this cockroach 'milk'. Unsurprisingly, it's not by attaching tiny little milking machines to the insects, but by using genetic modification to insert cockroach genes into yeasts that can then be bred up to make the 'milk' in a lab.

This kind of science reminds me of a wonderful story by John Kennedy, a British climate scientist, who posted on what was once called Twitter:

When an article says 'some scientists think' then remember this: I, a scientist, once thought I could fit a whole orange in my mouth. I could, it turns out, get it in there, but I hadn't given sufficient thought to the reverse operation.

This makes me laugh. As the son of a research chemist who thought science could solve all of humanity's ills, I can picture this happening. So often, from the PFAS forever chemicals that we have poisoned our people and cities with, the asbestos we pumped into roof cavities, to our enthusiastic burning of fossil fuels or even the discovery of trans fats to put in margarine, we get very excited about 'can we do blah blah blah' without ever truly considering the reverse operation. We don't ever quite seem to be able to predict the consequences of our actions.

The thing is, all milk – plant, animal and lab-made – comes with a consequence or three. Sometimes when you hear about agriculture, particularly animal agriculture, the numbers seem scary. The land use, the water use, the greenhouse gases. But we still need to eat, so if milk is to be replaced in part, or in whole, by alternatives, then we need to consider what that means in the whole system, not just at the factory floor. What happens to the dead yeast, for instance, when you scale up cockroach milk to industrial levels? Anything in quantity is problematic to dispose of.

For the last few hundred years, humanity's solution to pollution has been dilution. Let fresh water flush underneath a salmon farm to help dilute the high nutrient load from their poo. Flush a cannery's waste down a river. Throw food waste in a tip with all our plastic rubbish, cover it with dirt and walk away, fingers crossed. Or just spread the waste around and hope nature will fix it. But those days are coming to an end.

Along with waste, there are other consequences. Making lab-based milk is energy intensive. The yeasts need to be bred up at a certain temperature, and then they need to be dried, which consumes vast amounts of electricity. Is it renewable energy or not? The yeasts also need to be fed something. Is it nitrogen captured using fossil fuels? Is it palm oil? The microbes also need water. They need equipment. The factories need to use land and resources and human labour. And there's the waste. Always there's waste. While the data are sparse on the first few of these, we've been dealing with waste from plant-based milks for a while now, and it throws up some interesting conundrums.

Take oat milk, for instance. For every litre of oat milk you drink, there's about 200 grams (7 oz) of oat residue – oats that still have much of their fibre and some protein in them.[3] That's an awful lot of oat pulp kicking around.

Numbers for global oat milk production are tricky to find, as most companies keep them close to their chests, but it's been estimated that it could be as high as 2 billion litres (440 million gallons).[4] That would mean there's about 400 million tonnes of oat pulp to deal with every year.

The Sweden-based firm Oatly knows waste oats are a problem. They also know their big vegan audience became very upset when they found out where all that waste was going – namely, into pigs. At least they're very open about how they're trying to deal with oat waste:

Turning the oat residues into food for humans would be the most sustainable alternative, followed by animal feed, biogas and soil improvement.[5]

They say over half of their 41,000 tonnes or so of annual waste goes to energy use (about 58%), being used to make biogas, while 37% is fed to animals, according to data from 2021. And when they say animals, those are mostly pigs. If you're vegan and you think that's bad, by the time they put out their 2022 report, only 31% of waste was used for energy – and 59% (over 49,000 tonnes) was fed to pigs.[6]

Oatly's predicament is not unusual; waste is an eternal dilemma for food producers. What to do with the part of our crops that humans can't, don't, or won't, eat?

Traditionally, since Western-style agriculture started about 10,000 years ago, we've fed those wastes to animals, then milked or eaten those animals. It's called upcycling, and is often the most energy efficient and sensible use of a resource. When I asked several other oat milk companies what they do with their oat waste, some, like Germany's Blue Farm, said they feed it to pigs. Companies such as Rebel Kitchen, which ships its 'mylk' products all the way from the UK to Australia, and big player Bonsoy, said nothing. No response. Which says it all, probably.

Just about all plant milks end up with the same problem: a by-product that nobody wants to eat, or can't eat. One of these waste products has such a long history, it even has its own moniker.

Okara is the name given to the soy pulp that is left over after making soy milk. It contains little digestible sugar and is mostly fibre – namely cellulose, hemicellulose and lignin, which are virtually indigestible by humans. As with oat milk, for every litre of soy milk made, you end up with about 200 grams of waste – so 20% by weight. And as with oats, there's some goodness left in that residue, though it can be hard to extract. Okara has a pretty high protein component at about 25%, and it still contains 10–15% oil.[7]

So how much okara does the world produce? In China alone, they are left with 2.8 million tonnes of okara a year.[8] And that's only a fraction of the world's output, which comes in at a staggering 1.4 *billion* tonnes per year.[9] Much of this is fed to animals. Some is simply thrown in the bin.

All of this waste can be dried, to store and transport it better. It can be processed into other foods for people to eat – but currently not much is.

Oatly is spot on: the best use of this waste is to find a human food to put it in. But that food will, of course, also be ultra-processed by definition. You'd be hard-pressed to find nice home cooks willing to take some of that 1.4 billion tonnes of okara a year. Where this waste will end up is in factories. That could be why so many of the world's biggest food ultra-processors are getting into the plant milk game. More feedstock from a new source.

The second most efficient use of this waste from plant milks, after feeding it to humans, is to feed it to livestock. About 85% of what is fed to livestock (when the animals aren't ranging on grass) is unsuitable for human consumption – think straw from barley, grains that are cracked and broken or otherwise damaged during harvest, spent grain from brewing, that kind of thing. Plant milk residue is usually ideal, because it often contains relatively little fast-digesting sugar (which is bad for ruminants), but still contains fibre and protein that, depending on the ratios, some ruminants – and all pigs – thrive on.

So, oats and soy are big producers of waste. You'd think almonds might have a better fate? And you'd be wrong. When you take an almond off the tree, most of the weight is the hull, which is nutrient dense, but not digestible by us. Only 11% is the actual usable nut. If we waste 89% of the almond, have we done the right thing by nature?

Well, the good news is that we don't waste them. The shells are used for chicken and cow bedding, and sometimes in their feed. The hulls (the nutrient-dense outermost part) are used to feed dairy cows and chickens.[10] Almond hulls are also used to grow black soldier fly larvae (the maggots we met earlier that are being made into insect milk), but those larvae are usually added to the feed of fish and poultry farms, not put in the mouths of humans.[11]

Almond pulp, the waste left after you strain off the milk, is used to grow mushrooms, which sounds vegan friendly, but it seems to work best only after it is mixed with the waste from chicken sheds.[12] The hash – the bits of nut that fall off after de-shelling – are used as a high-end feedstock for cattle.

In Australia, where large almond farms have found a home because of water shortages in California and a ready ability to buy water rights here, some almond waste has been simply left to rot, returning its carbon dioxide to the atmosphere (or, if piled up, releasing methane). A lot also used to be fed to farm animals, but as the price went down, it's not even worth shipping to feedlots anymore. Most is now apparently burned and used as a soil amendment, biochar.[13] And while I love the prospects of biochar (I devote a whole chapter of my book *Soil* to it), in the order of waste's preferred use, it's not the first or even second-best option.

Almonds have the dubious reputation for the highest water use of all plant milks, with US estimates at about 12 litres (2½ gallons) of water per individual almond,[14] and a similar amount in Australia. According to the Sustainable Restaurant Association, it takes over 6000 litres (1320 gallons) to produce 1 litre of almond milk.[15]

As we've seen with real dairy, this isn't a problem if the water falls from the sky, but could be a problem if it has to be pumped from a dam. Just to put it in context, Californian dairy, which is usually produced indoors from cows fed irrigated grain and silage, uses 1.8 times the water of almonds to produce a litre of milk.[16] To produce a litre of grass-fed dairy milk, such as we mostly see in Australia, it takes roughly 743 litres (163 gallons) of green water (rainfall) and 44 litres (10 gallons) of blue water (from reservoirs, bores or dams) – though, as always, individual farms vary.[17]

In Australia each year, we produce about 120,000 tonnes of almond kernels[18] (requiring about 200,000 bee hives to pollinate the trees[19]). If all these kernels were made into almond milk, it would leave 3600 tonnes of dry pulp, and way more in terms of wet matter, which would need to be dried – a process that is very energy intensive.

These figures are dwarfed by the US, where they produce over 1 million tonnes of almonds, which would leave at least 30,000 tonnes of dry pulp. Far smaller than their oat and soy counterparts, for sure – but remember, nearly 90% of the almond weight was in the hull, which has already been stripped off and fed to animals.

And then there's the bees. In California, which produces the vast majority of the world's almonds, some estimates put the bee death toll within the industry at up to 40 billion annually. Australian numbers are harder to source, but if the US experience is anything to go by, where half the bees die, that would mean the death of about 5 billion bees here from the 200,000 hives used in pollination each year.

Animal agriculture is agriculture. It sits at the nexus of soil, plant and human ecology. Using animals to consume biomass is a good way to upcycle nutrients. I'm not saying it's the only way, but it's one way. And you'd have to be an idiot, and want to insult Mother Earth, to not use her resources wisely.

Sitting in a café in Greenwich or Melbourne or Toronto, it may seem weird that the plant milk in your coffee may in fact support animal agriculture further down the chain, but ecological necessity and financial practicality (and those on the ground actually growing your food) show that is the case. Already, some estimate that 38% of American-grown grains go to landfill.[20] If it's in landfill, it creates methane.

There are, however, clever people coming up with other ways to reduce or use the waste. At least one company, Willa's in the US, has worked out how to use more complex enzymes to help ensure that most of the whole oat is used in their oat milk. This seems very wise to me. Virtually no waste, and all the nutrition going to the drinker.

Other companies are drying the residues – a process that is energy intensive, so only good if the electricity used is renewable. Elajo, a processing company in Sweden, is working on the technology to produce dried oatmeal for use in commercial food production. They're also looking at extracting the protein to use in meat substitutes, though that is still in the lab stage. Dried pulp can be used in the industrial food system, though again, much ends up in the mouths of animals. Some companies, like Califia Farms, well known for almond milk in the US, send their pulp to a waste contractor that makes slurry. This ends up on soil, so it's not a bad end, but it'd be so much better if it was further up the food chain.

Looking at those lab milks, the ones using genetically modified yeasts, they're still at such a small level that their own waste problems haven't scaled up much yet. If their operations increase greatly in volume, however, all those dead yeasts from their lab fermentations will also need to be upcycled. According to Professor Paul Wood from Melbourne's Monash University, to get 10 grams of protein, you need about a litre of fermenting yeast.[21] In other words, you'd need three times as much yeast culture as milk to get the same amount of protein. That's a lot of yeast. Scaling that up, in Australia alone, to replace dairy, we'd produce 26 billion litres of yeast fermenting waste every year.[22] What does 26 billion litres look like? Imagine 14,000 Olympic-sized swimming pools.

Animal proponents often say it's not the cow, it's the how – meaning it's not the animal, it's the way it's farmed and the way its product is treated. But in fact, it's not the oat or the almond or the cow. It's the *how*.

Comparisons are really hard to make to say one thing is saintly, the other not. Oat milk, for instance, is responsible for about 50 times more phosphorus use than goat's milk, thanks to fertiliser use. One almond milk takes 24 times more energy to produce than another. And one system for producing real dairy milk can use up to 13 times more energy than another.

While it's relatively easy to find numbers on dairy cows and their emissions and waste, little is actually known about the broader impact of new milks. There's been very few life cycle assessments done on cashew, coconut, hazelnut, hemp, peanut, quinoa, sesame, tiger nut and walnut milks.

In a 2022 environmental and nutritional review of milk substitutes, researchers found:

> **Even on the available data for the other products (cow, buffalo, goat, sheep, oat, almond, rice, and soy), there is missing data that does not allow to make a complete comparison of the available data. The only category that has data available for the mentioned beverages is global warming potential. This is an important category for the life cycle assessment, however, a comparison of products or services in this sole category does not show the real impact of the product.[23]**

All of this should be seen in context. Is plant and lab milk sustainable? Perhaps, depending on the measure. It's worth remembering that just as everything has a physical consequence, it also has an economic consequence. Fake milk is sucking a lot of money out of improving agriculture while we work out if cell factories and plant milks can actually be scaled up sustainably.

And in the meantime, some plant milk isn't popular enough to make it profitable in its own right. Much is doing well only because of investors hoping to cash in.

Oatly, the world's biggest oat milk player, saw their share price drop 90% at one point in 2022. In 2023, their sales were up 12%, while at the same time losses were up 87%, with the company losing a third of a billion dollars in a year. For every $1 of oat milk they sell, they spend $1.48. They don't make a profit in any market in which they sell.[24] They even had to withdraw their dairy-free 'ice cream' from British shelves in 2023 due to lack of demand.[25] Perhaps it's just an aberration as they expand. Or their trajectory could be similar to those of meat alternative companies such as Beyond Meat. In July 2019, Beyond Meat's share price peaked at US$234.90, but at the time of writing this book, it sat around US$13. The hype about fake meat has been met with excitement, then reticence, then commercial reality. It's unlikely it will put real meat out of business – perhaps because it's not actually that nice, or nutritious, or as low impact as some would have you believe.

Will fake milk have the same initial disruption thanks to media hype and plenty of advertising spend, then later plateau? The take-up by consumers seems greater, and the products themselves mimic the real thing more readily; they can be made to look and perform, in very superficial terms at least, like real milk.

There's an old farmer saying: it's hard to be green when you're in the red. Oatly seems to be trying to be green while haemorrhaging money. That's how big business works when there's cheap money flowing in from speculators, but is it how we feed the world?

I think fake milk's problem isn't what it *is*, but what it's trying to be. Because while plant and cell factory milk are pretending to be milk – hence the addition of vegetable fats and emulsifiers and calcium and more – they aren't milk in very fundamental ways. To really understand why, we need to look more closely at protein. Not just the *quantity* of protein, but also its quality.

Protein Shake

You don't want to milk a cranky camel. Their milk letdown takes way longer than a cow's, and requires a vigorous udder massage. And where a goat's kick is minor, a sheep's leg can barely cause a flesh wound, and a cow's kick can kill you (but their legs have a very limited range), a camel can flick its wide, hard hoof in just about any direction it likes. One dairy owner calls it a 'taekwondo kick'. And if a camel wants to lash out, kicked you'll be.

People are chancing it, though. Camel milk is the latest in a long line of animal milks to make it to the mainstream. Along with the usual suspects, humans also milk elk and yaks – and someone somewhere is even drawing milk from pigs and making cheese. And where the climate suits, some people are milking camels.

What makes camel milk stand out from the herd is not just its novelty, but its protein.

Different animal milks vary in their protein amounts and composition. Cotton-tail rabbits, for instance, have milk that's nearly 16% protein – about five times that of commercial cow's milk. Human milk has one of the lowest protein levels of all, less than 1%, which some think helps explain why it takes so long for our babies to mature. Rat milk has 10 times more protein than ours.

Camel's milk differs from human breastmilk and cow's milk in a more fundamental way. It lacks beta-lactoglobulin, which is a major protein – and also a major allergen – in cow's milk.[1] However, the milk of the desert dweller does contain microbial agents such as lactoferrin, lysozyme, immunoglobulin and lactoperoxidase, which all have an overwhelmingly positive role in fighting bacterial infections in our bodies. (We've met some before, and I'll get back to those other sciencey-sounding things in a bit.)

I'm going to go into some depth in these next two chapters. I need to because milk is so complex, and to understand what it does well, we need to look at both the complexity of what is in milk, and the intricate relationship it has with our bodies. I'll try to highlight the important bits and let you know which bits you can skim read.

Now, a proviso. The following information is broad in nature, based on populations or groups, not on some miracle of milk to cure everything in every individual, so don't look at it as medical advice. I'm trying to find out what it is about this mammalian superfood that has had us not only consume the milk from other animals for 10,000 years, but also have genetic mutations widely circulating in society that allow some of us to absorb all milk's nutrition.

Protein in milk can be broadly broken into three groups: caseins, whey proteins and mucins.

Caseins make up about 80% of cow's milk proteins, and whey proteins about 20%. By comparison, in human breastmilk the protein mix is about 40% caseins and 60% whey proteins, which is one of the reasons we need to adjust cow's milk for infant formula. Mucins, which are best described as proteins that exist as membranes around fat globules, account for only about 1% of proteins in most animals – but that 1% has an important role, too.

I'll take a look at mucins first, because they're the simplest. Mucins line things. They share the origin of their name with mucus. While that's enough to put some people off milk, scientists think that their ability to form a lining – a mucus layer, essentially – means mucins can act as an anti-cancer agent, and perhaps is the reason why dairy cows very rarely get mammary tumours, their version of breast cancer.[2]

Mucins are notoriously hard to isolate from milk, and only a couple have been identified in human breastmilk. When we drink milk, whether human or cow, the mucins line the cells of our mouth and gut. The mucins in human breastmilk have been shown to reduce the susceptibility of cells to attack by pathogens such as the AIDS virus (HIV), rotavirus, *E. coli* and salmonella.[3] This quartet are real baddies, so if mucins can stop these attaching to the inside of our bodies, that would seem like a really good thing.

Mucins in cow's milk are slightly different to those in human breastmilk, but line the gut in a similar way. Research has identified three different mucins in cow's milk, and they probably have a role similar to those in human (and rodent) milk, including reducing the activity of dangerous bugs such as *Helicobacter pylori*, the bacterium that causes peptic ulcer.[4] They also seem to be particularly helpful in lowering the incidence of gut and respiratory infections. Remarkably, cow's milk mucins have also been shown to inhibit infection with Covid-19.[5] Milk is not a Covid vaccine, but the mucins have been shown to help in preventing initial infection. (So, too, lactoferrin.[6] There's also research into milk oligosaccharides and their usefulness in long Covid.[7])

So, those mucins represent a small but really important 1% of cow milk proteins.

What about those caseins? They're the major proteins in ruminant milk – the sort that are being made outside of actual cows using genetically modified yeasts in labs. Representing about 80% of proteins in cow's milk, caseins do some pretty noteworthy things. The first is that they are naturally quite soluble, and hang on to calcium reasonably well. This means milk doesn't separate out into some kind of sludge at the bottom and water in the middle and all the fat at the top. Importantly, it also means the calcium in milk is in a good form to be digested.

There are different forms of casein, and depending on what research you want to look up, they could be named alpha, beta and kappa. Goats have at least 64 different caseins in the alpha, beta and kappa range. In cow's milk, there are at least 13 forms of beta-casein alone.[8] (To further complicate things, alpha-casein can be alpha-s1 or alpha-s2. Alpha-s2 is called epsilon-casein in mice, gamma-casein in rats and casein-A in guinea pig milk!)

I told you it was confusing.

Anyway, let's just make it simple. There are lots of caseins, and they have weird names. Caseins give real dairy its texture. In cheese they allow the curds to clump together and knit, and they help yoghurt set, too.

In our bodies, caseins have been shown to reduce blood pressure, increase the activity of immune cells (in this case, macrophages), increase white blood cells (innate immunity), help make antibodies and cytokines (both of which are essential to our immune systems), and seem to suppress cancer cell proliferation.[9] They can also aid in muscle repair, give feelings of satiety (fullness, so they reduce calorie consumption), aid the body's response to vaccines, and have been shown to improve athletic performance. They're generally considered beneficial in cardiovascular disease. So, unless you're allergic to them, it's likely that caseins are helpful healthwise. Just remember, while these are all effects we've witnessed, they're quite specific to one study or another. Decreasing blood pressure using caseins, for instance, was more effective on men of Japanese ancestry, for reasons as yet unknown.

Caseins get broken down in our bodies into peptides. And peptides do all sorts of good things for our bodies, too. I'll circle back to peptides in a minute.

The third major type of protein in milk after mucins and caseins is whey proteins. These represent just a bit under 20% of the proteins in cow's milk, and about 60% of the proteins in human breastmilk. They're called whey proteins because if you coagulate milk to make cheese, the caseins set into curds and the whey proteins fall out in a watery fluid. Curds and whey.

Whey has five major protein groups that make up about 85% of their mass, and I'll focus on three.[10]

The first protein I want to highlight is alpha-lactalbumin – which, as we saw before, makes up most of the protein in human colostrum.

This molecule helps form breastmilk's HAMLET protein we met earlier. (HAMLET, if you recall, being shorthand for Human Alpha-lactalbumin Made LEthal to Tumor cells.)

HAMLET does some pretty amazing things. It doesn't just modulate cancer cell growth, it actively kills tumour cells, making it a thing called a tumoricide. It is also a bactericide (bacteria killer) – particularly the streptococci strains, which are responsible for strep throat, impetigo and rheumatic fever, among other illnesses. Even more than that, HAMLET also helps increase the ability of conventional antibiotics to do their job. For instance, erythromycin-resistant bacteria can be killed with erythromycin far more readily in the presence of HAMLET. It pimps antibiotics, to help fight superbugs.[11]

Alpha-lactalbumin is present in smaller amounts in cow's milk than human milk, but it's still important, being the second most abundant whey protein, at about 20% of the total. Alpha-lactalbumin increases the absorption of iron in the body. It can help decrease blood pressure, reduce stress, and is anti-inflammatory. It also has a role in neurological (brain and nerve) functioning, and can be an aid to better sleep. When it comes to enhancing immunity, alpha-lactalbumin has been shown to be more effective than other proteins, such those from soy, or wheat, or even casein.[12]

And while alpha-lactalbumin on its own is good, it's better when combined with other compounds, just as we saw with HAMLET. We're just learning about a thing called BAMLET, which stands for, you may have guessed it, Bovine Alpha-lactalbumin Made LEthal to Tumor cells. So, cow's milk alpha-lactalbumin also kills cancer cells, even if its name isn't quite so Shakespearean.[13]

Did I tell you how amazing alpha-lactalbumin was?

The second whey protein I want to highlight is beta-lactoglobulin. This major protein of cow's milk seems to be absent in breastmilk as well as camel milk. Does that mean we should be ingesting it at all, as adult humans?

In cow's milk, beta-lactoglobulin represents about 65% of the total whey proteins (about three times more than alpha-lactalbumin). While beta-lactoglobulin was first isolated in 1934, it's still a bit of a mystery, according to the boffins.[14] We do know it boosts immune responses, as alpha-lactalbumin does. Researchers have also found it has properties, tagged BLAGLET (you can work it out), where it has been shown to be lethal to over 50 different cancer cell lines, in test tubes at least. And beta-lactoglobulin also has the capacity to lower blood pressure, and lower blood serum cholesterol.[15] I'll talk more about cholesterol and heart disease in the next chapter.

It's not all rosy, however. Beta-lactoglobulin, just like alpha-lactalbumin and casein, is potentially a milk allergen, particularly to young infants. This isn't an intolerance, like lactose intolerance. It's an allergic reaction, mainly in kids. The allergy only affects about 2% of infants in developed countries, and far fewer adults, but it can be serious, and the proportion of those affected is rising.

If you're not allergic, though, it seems that both the alpha and beta whey proteins do really good things for the body – particularly when they work in concert, as they do in whole foods, not just in isolation or as additives.

Milk proteins are often broken down in the body to form peptides. These protein fragments can do all kinds of things for the body, from having opioid-like features (painkillers and mood lifters), to helping calcium uptake, to being able to stimulate the immune system, to being anti-hypertensive (lowering blood pressure), antiviral and antibacterial.[16]

When we digest casein, one particular peptide we end up with is known as glycomacropeptide, which also forms during cheesemaking.

This single protein is antibacterial. It acts as a prebiotic, feeding our microbiome. It can remineralise our teeth, is considered anti-tumoral (cancer cell fighting), and it moderates our immune system as well as the process of digestion.[17] In other words, it helps fight disease by killing bugs, by boosting the immune system, by reducing the chance of cancer cells regenerating, and by feeding the good bacteria in our gut. Not bad for something that's hard to say.

One of the most exciting discoveries in the milk of many mammals is lactoferrin – an amazing chemical that is thankfully easy to pronounce. We met it earlier, in the Colostrum chapter and elsewhere, but it's not just me who thinks of lactoferrin as having some kind of superpower. In a 2022 article in the journal *Molecules*, titled 'The Lactoferrin Phenomenon – A Miracle Molecule', the authors note it has:

> **beneficial properties such as anti-pathogenic, anti-cancer, anti-inflammatory, immunomodulatory and DNA-regulatory activities. Recent reports indicate its therapeutic properties in the treatment of neurodegenerative diseases associated with aging, as well as stress-related emotional disorders.**[18]

We've done a lot of research on lactoferrin – from the lab, to isolates in the diet, to how it works in milk. It's a cracking performer. Lactoferrin helps bone development and helps reduce bone breakdown, even reducing the progression of osteoporosis in old age. It helps the gut flora in our own microbiome get back to normal after a course of antibiotics. It helps us absorb dietary vitamin D, is antiviral, and also assists antiviral drugs to work better. It helps cells avoid being colonised by baddies such as the Covid-19 virus. Being an anti-inflammatory and antioxidant, lactoferrin is considered to have anti-ageing properties. It improves glucose metabolism in people with type 2 diabetes. It regulates body fat metabolism, and can reduce blood pressure and help in cardiovascular disease.

Consuming milk is no guarantee of health, or of recovery from disease, of course, but lactoferrin sounds like something you want on your side. And if you can get it in your diet, it's probably not a bad thing.

Add to that the immunoglobulins – essentially antibodies – that are also a major fraction of milk protein, and you've got a veritable soup of goodies in a glass.

If all that seems too good to be true, don't forget that milk is designed to do things no other food does. Its role is not just to nourish, but to fortify, to communicate, to build the robustness of the animal that drinks it.

And what I'm about to tell you next makes all those lacto thingies and peptides and HAMLETs look like yesterday's science. There are things in milk that alter your very DNA.

Our DNA is our genetic code. We used to think it ran the show, deciding what biochemicals our body should make, what cells to produce, depending on the genes embedded in it. Our DNA determines our eye colour, our height, our blood type, our likelihood of getting heart disease.

We now know that what DNA does is also affected by other things – including the non-human genetic code in our microbiome. How genes are expressed depends not just on our gut flora and vice versa, but also on what we eat.

We now also know that things in milk, called 'micro RNA', written as miRNA, can alter not only the way we build and protect our bodies, but also our bodies themselves. This happens not just in our guts, but in the very periphery of our bodies, such as the nerve endings on the tips of our fingers. According to scientists, 'miRNAs in milk are bioactive food compounds that regulate human genes'.[19]

Now, some miRNA can exist in plants such as broccoli, too. But researchers have found that they exist in smaller amounts than in milk, and that broccoli miRNA has no discernible effect in humans, as it is probably broken down or removed during the process of digestion. And while lots of plants do contain miRNA, scientists point out that 'previous claims by a single laboratory that miRNAs from plants affect human gene expression are highly controversial and have largely been dismissed by the scientific community'.[20]

Dairy miRNA, on the other hand, is protected on its journey into our blood by the other things in milk. Once there, the human vascular system – our blood vessels – are lined with cells that can take in milk's miRNAs, and move them around the body. The miRNAs from milk can be found in the spleen, heart, liver and brain in laboratory animals, so we presume the same happens in humans – that they get into our bloodstream and into our organs. And when I use the plural miRNAs, we're talking about a *lot* of them.

So far, about 1400 miRNAs have been identified in human breastmilk,[21] and 1025 miRNAs[22] in cow's milk.

These tiny fractions of genetic potential are probably made by the mammary gland, but some are made and exuded by other cells. You find them in skimmed cow's milk, but there's more in whole milk. And looking at the milk you put on your cereal, 89% of the miRNA found in human milk is identical to that found in cow's milk, and 83% is the same as in goat milk.

There are hundreds of miRNAs in every mammalian milk, and nature didn't put them there by accident. Something from a mother, or a cow, that can affect the expression of our DNA, enters our brains and affects muscle and bone development. And from all the research so far, these nanoparticles seem to have overwhelmingly positive effects on body tissues, immunity and neurological health – even if we drink milk from a species not our own.

MicroRNA usually comes delivered in an EV – an extracellular vesicle, something that is put out by cells. EVs carry proteins, fats and metabolites. They play a role in immunity and in the general running of our bodies. Milk EVs more broadly have been shown to do all kind of things from immune regulation to suppression of arthritis.[23] Once thought of as cellular waste, EVs are now known to influence all kinds of bodily functions.

In many respects, EVs provide much of the unspoken communication between a mother and her child at the cellular level.[24] There's information being passed between generations through breastmilk that has an impact on how our genes are expressed, altering not just those genes for the next generation, but for all the ones following. So, a mother's breastmilk affects how her child's genes might express themselves. It's called epigenetics, and all milks can definitely play a role. Certainly, breastmilk has a major role, starting on that very first day of life outside the body, when our young gut is more pliable, our tissues still maturing – but it's believed other mammalian milk we drink can impact us all through life, too.

Why am I telling you all this? Because I want you to understand just some of the complexity of milk, some of the ground-breaking research that's still being uncovered, and to realise that milk is the result of a lot of evolutionary heavy lifting. It's a miracle liquid governing everything from our feelings of satisfaction to the development of the nerve endings in our fingers. Stuff in milk can not only taste quite nice, but can also help cells communicate – and even alter the way our genetic code is expressed. It aids in the fight against cancer, and feeds our own internal rainforest of microflora, all while being able to help you sleep.

Now, all of this should be seen in context. There are anti-cancer and immune-boosting agents in fruit, vegetables and nuts. And scientists are often trying to isolate fractions of milk (or blueberries, or quinoa) and put them in pills. Or formula. Or protein drinks for gym junkies.

But nothing is as complex as milk. It doesn't just have one or two of these chemical compounds, it's packed with them. Remember, there are 4654 proteins in cow's milk that we've identified, so far. Many may perhaps be benign for a grown human compared to a young calf, lamb or kid, but lots of them are doing you immeasurable benefit. You don't have to drink milk, or like milk, or think it's a valid food choice for you, or your adult family, or the planet – but you have to admit, it's pretty damned special.

Does that mean you should drink more of it? Or a lot of it? Well, no. Just as there's vitamin B12 and microbes in soil that can inoculate your gut and help release happy hormones in the brain, and yet you shouldn't eat a half cup of soil a day, it's the same with milk. Some is good; too much might be less than good.

Balance is the key. Sure, milk sugars play a role. Proteins seem pretty vital, and peptides, microRNA and other extracellular vesicles seem mightily important, too. But there's also another source of dietary goodness and potential badness in dairy. We need to talk about fat.

You Are What Your Cow Ate

Fur seal mums would be quite happy with Dr Emmett Holt's baby feeding routine, which we came across in the chapter Formulaic. In a human mother, limiting night feeds to a single suckling can put her milk at risk of dropping so low that she has trouble producing enough to nurse her baby. But that's not true for a fur seal. She can go for up to four weeks without suckling her child – a feat not seen in any other animal.

A fur seal mum can do this because her milk is up to 60% fat, which provides 85% of the infant's energy needs. This allows her to go foraging for food for weeks at a time before swimming back to shore to discover that her baby hasn't died of hunger.

Milk macronutrients – the fat, protein and sugars – vary a lot between species. Animals generally have one or two at higher levels. No animal can make ice cream, which is high in all of them.

Human milk is relatively low in protein, high in sugar, and not dissimilar to a cow in its fat ratio, at about 4%. It's considered quite dilute in the mammalian world. That means we need to feed our babies more often, and for longer, than animals with a different milk composition. Cow's milk, for instance, is less dilute because it's higher in protein than a human's, with natural weaning occurring at about 10 months.

Meanwhile, the actual fat *composition*, not just the total percentage of fat, varies a lot between species, and also according to the stage of lactation, the baby's development, and the mother's diet, genetics and environment. If you're drinking milk, you're not just consuming one type of fat, but several, which also depends on what the cow (or buffalo, or sheep, or your mother) ate. Or what kind of fat they've put into formula or fake milk.

If you drink milk, you're being altered from the inside out by what that lactating animal ate.

I was waiting for the supermarket to open. Outside, a rotating electronic billboard showed an advertisement for butter, then, during its cycle, ads for two brands of cheese. Oh, and look, there's an advertisement for thickened cream:

> '**Ah … turns out we didn't need the digestive system of cows after all. Skip the cow, not the taste.**'

Flora's 'thickened plant cream' is made from lentil protein, coconut and canola oils, sugar esters of fatty acids (which are used in detergents and as emulsifiers in cosmetics and the agricultural chemical industry), along with another 12 or so ingredients that can pretty much only come from a factory. Some are other fatty acids, and there's lecithin from sunflower oil production.

Real cream, the old-fashioned sort that famed margarine maker Flora is replacing with 'plant cream', is about 35% butterfat. The rest is made up of the milky bits we saw earlier – the proteins, milk sugars, and their associated EVs and miRNA. This butterfat can be churned out of the cream, to make butter. Butter itself is about 80–85% fat (along with a small amount of milk solids and some water).

And while butter may look like a singular product, it's not a single fat, but rather multiple fats: short-chain and long-chain fatty acids, and saturated, polyunsaturated and monounsaturated fats. So, let's do a quick Fats 101 to make sense of what I'm talking about.

Fats 101

Fats are made up of strands called fatty acids. These fatty acids can be saturated or unsaturated. And unsaturated fatty acids can be monounsaturated or polyunsaturated.

Righto, so, saturated fats? Well, we used to think they cause heart disease, but that idea has pretty much been debunked. Are they healthier than some other fats? Hmm, perhaps not. The general consensus is to not overdo them.

Monounsaturated fats? Pretty good, actually. Generally they're from seeds or nuts, with olives a good source.

In the polyunsaturated fat camp lie a few options. Unsaturated fats can be omega-3, omega-6 or omega-9. Don't worry about why they're called that. Just know that all three are essential for us to function effectively – for the development of all our cells, including those in our brain and internal organs. Our bodies can make omega-9 fats, so although it's good to have some in your diet, we can synthesise them from other sources. Our bodies cannot make omega-3 and omega-6 fats, however, which means we absolutely need some of these in our diet to properly fuel our bodies.

Humans were designed for a nice balance of these fats. We evolved to eat them in a ratio of about 1:1 omega-6 to omega-3, or a maximum of 2:1. This ratio is good for brain development, heart health and more. The problem is, most people in developed nations eat a bad balance of these fats. In the US and Australia, the ratio is about 20:1. Omega-6 fats are necessary for us to function, but we're eating 20 times more of them than omega-3 fats. This enormous increase is because we now eat more plant oils – both in the processed food we buy, and in the oils we choose to use at home. And it's not good news for our bodies because they cause inflammation when consumed in excess.

Omega-3s are generally considered really beneficial in the diet, especially if eaten in the right proportion to the other fats, in particular omega-6. One suggestion is that if you eat more omega-3s, and less saturated fat, you lower your chance of cardiovascular disease, but the research isn't conclusive. Omega-3s come in a few shapes and sizes, which complicates things a bit, as the type found in seaweed (and hence, fish) is slightly different – and slightly more beneficial – to the kind in land plants (and hence, milk).

Omega-6s are a bit of a minefield. On one hand, they are vital for brain function and normal development. On the other, there is some evidence that some kinds of omega-6s play a role in increasing inflammation in the body, and can be bad for heart health and also contribute to degenerative diseases. Overall, though, they're nowhere near as bad as plant trans fats – factory-made plant-based fats that were once very common in margarine and food processing, and still exist in some foods in some jurisdictions.

Some omega-6s are clearly beneficial. Conjugated linoleic acid (CLA) is an omega-6 fat that most people get from ruminant animals (by consuming dairy and meat), and it has been shown to have really good health benefits, which are quite different to the omega-6s found in grains. CLA has been shown to be a potent anti-carcinogen, as well as having anti-atherogenic (heart-disease fighting), immune-modulating, anti-diabetic, cholesterol-lowering properties.[1]

CLA is also made in our bodies when we consume trans vaccenic acid (TVA), an animal trans fat that comes mostly from ruminant milk and meat. Consuming animal trans fats is completely different from plant trans fats. TVA, for instance, reduces the incidence of heart disease and obesity, as well as slowing the rate of several types of cancer, and helping improve other cancer treatments.[2]

Fats ain't fats

What happens in the body is what counts. In the old days, many of us were warned about cholesterol, and were told that increased blood cholesterol was bad. But it turns out it's more complicated than that, sigh. There's both low-density lipoprotein (LDL) and high-density lipoprotein (HDL) cholesterol. And one is better than the other. More HDL is good, more LDL is bad.

We used to think dietary cholesterol was bad. Now we know it makes no difference.

Then we used to think that *all* saturated fats were inherently bad, including dairy fats. Now we know that fermented dairy products such as yoghurt and kefir either don't raise LDL cholesterol, or can help reduce it, depending on the study.[3] And full-fat dairy (the kind banned in US high schools) does a better job of increasing the good HDL compared to skim milk.[4]

Dairy fat – broadly speaking – is either good for your heart, or perhaps neither good nor bad. It's certainly not bad.

Let me tell you, fats are complicated.

Okay, so back to spreading some thick, rich butter on the topic.

Butter is roughly 66% saturated fat, 26% monounsaturated and 3% polyunsaturated, with a few other minor fats thrown in. But these numbers are rubbery, and they depend a bit on the cow, and a lot on the how. Does this mean butter is bad for us? Well, not necessarily. Butter actually has a nice combination of fats that seem to work well together, but what your cow ate changes the ratios, sometimes in ways that aren't very good.

Organically raised cows that eat pretty much only grass can have 7% more polyunsaturated fat in their milk than non-organic milk, and slightly less saturated fat (which, broadly speaking, is a good thing). And organic milk has 56% (yes, a whopping 56%) more omega-3s than non-organic milk.[5]

I'm going to sound like a broken record on the good things in milk, but fats do have a role in a healthy diet. Remember the CLA that you get from ruminant milk? It's an omega-6, but a good version.

One study found that CLA 'is the only fatty acid shown unequivocally to inhibit carcinogenesis' (the initiation of cancer) in experimental animals. It's also good for bone formation, preventing heart disease, treating and preventing diabetes, and the list goes on.[6]

If only we knew what the cow's diet does to the CLA in milk. Oh, that's right, we do. It turns out that grass-fed cows have more CLA in their milk than their counterparts in sheds.

In one study, CLA in milk was increased between 60% and 99% in cows fed forage (grasses, essentially) rather than grain and hay. They also had 30% more of other beneficial fatty acids (such as the omega-3, alpha-linolenic acid) in their milk. And their milk boasted up to a 50% increase in vitamin E, and up to 80% more antioxidants in the form of carotenoids.[7] In other words, more goodness. The milk of predominantly grass-fed cows contained less total omega-6, too, which fits the ideal ratios we're looking for.

A study from Slovakia showed you can increase CLA in butterfat by about 2–3 times if the cows are fed on grass, rather than in barns on dry fodder and grains.[8] In Wisconsin, they quadrupled the amount of CLA by putting animals on grass, not a grain mix.[9]

Some fats are formed independently of the diet of the animal. There's a group of fats called sphingolipids. Now, if you've never heard of them, you're in really good company, but these fats are pretty good at providing myelin sheaths for your nerve cells, so they're good for brain health. They're involved in regenerating all kinds of cells, help reduce inflammation and play a positive role in fighting cancer.[10] The human mammary gland makes these sphingolipids independently of what a mother eats, so they're present in breastmilk, but you don't get nearly as much from cow's milk, and very little – if any – in formula. Another reason it's good to breastfeed your baby if you can.

You can change a lot of the other fat content in milk depending on what you feed the mother. Feeding goats acacia (wattle seeds in particular, but also leaves) has been shown to improve the omega-6 to omega-3 ratio, and the fatty acid content more generally.[11] It's possible to increase the ratio of poly and monounsaturated fats in shedded dairy cows by feeding them soybeans or fish oil. But both of those have unintended consequences. Fish oil is taken from a very depleted resource, the ocean. Soybeans already contribute a lot to deforestation. (And the majority is turned into soybean oil for humans, which makes about 45% of the cash from soybean farms,[12] despite only comprising 20% of the bean[13] – and all the resulting waste, or soybean meal, is fed to livestock, mostly pigs and chickens.)

The good news is that we don't need to feed a cow soybeans or fish oils. A cow doesn't make its own omega-3s, but gets these from grass in the same way a fish gets them from seaweed. Omega-3s are used in photosynthesis, so any green leafy plant is the original source. Animals bioaccumulate them, meaning omega-3s are more nutritionally dense in the animal than they were in the plant. And studies have shown that you can get an omega-6 to omega-3 ratio down from nearly 15:1 in the milk from barn-reared, grain-fed dairy cows to 1:1 in fully pasture-raised animals. What your cow ate matters. While the 15:1 ratio seems to be an outlier, another study showed a ratio of omega-6 to omega-3 of about 5.8:1 for conventional milk, which dropped to 2.3:1 for organic milk, and 1:1 for the milk from fully grass-fed dairy cows.[14] The holy ratio of 1:1 *is* possible, depending on what your cow ate.

So, that's omega ratios and saturated fats and CLAs and more.

There are other fats, too, which is why milk is considered the most complex 'lipid' on Earth. The last fat I'll mention here is butyric acid. Milk fat is one of the very few food sources of butyric acid, which is a potent inhibitor of cancer cell proliferation, and actually programs cell death in several cancer cell lines.[15]

If these are the fats that are in real milk, what's in your plant-based mylk? Or your yeast-fermented lab milk?

Well, it depends. For a start, it could be a good oil, perhaps from a seed with a nice oil profile such as avocado oil, which is high in heart-healthy oleic acid,[16] but it's significantly more expensive, so I've not found a plant milk that uses it. Or it could be sunflower oil, with its disastrous omega-6 to 3 ratio of 40:1, or even 60:1. Remember, we want it as close to 1:1 as possible.

It could be canola (rapeseed) oil, which in the US is most likely to be from genetically modified plants. Canola oil's omega-6 to 3 ratio is a pretty good 2:1, but the oil has usually picked up a few dangerous trans fats from being heated to deodorise it – and the omega-3s are also degraded when heated to extract the oil from the seed. Cold-pressed is way better, but very little canola oil is cold-pressed.

The fat added to your plant milk could be palm oil, which is *the* major contributor to deforestation in South-East Asia, and has been shown to increase blood cholesterol in healthy people. Palm oil is about half saturated fat, and its omega-6 to 3 ratio is about 30:1[17] – some say over 40:1.[18]

And then there's coconut oil, which is about 80–90% saturated, and doesn't have the nice mix of other fats that butter has.

And all those oils take fossil fuels to grow, to fertilise, to harvest, to transport, to extract and then put into the new milk. Is that better for the world?

Milk, of course, isn't just fats and proteins and sugars. It also contains natural phytochemicals. By definition, these start in plants (*phyto* – meaning 'plant'), and there are thousands of them, many of them playing important roles in human health as antioxidants, poly-phenols and all kinds of things. And, many of these can be passed into milk. While plants are a better source of most phytochemicals (and I'm a huge fan of fresh vegetables), some can be more prevalent in milk than in broccoli. The thing is, the more diverse diet a grazing animal has, the greater their phytochemical load. In fact, there can be 20 times more phytochemicals in milk and cheese from an animal that has grazed, compared to one fed grain and hay and the like.[19]

A great example of a nutrient that is affected by diet is vitamin A – or beta-carotene, as the form we'll talk about is known. Beta-carotene (and carotenes more broadly) is a chemical in green plants that helps them photosynthesise, along with chlorophyl. Beta-carotene takes its name from carrots, because in its isolated form it can appear orange. When leaves turn orange or yellow in autumn, it's because the chlorophyll has gone and the carotene is left for a while before it, too, oxidises and the leaves brown off.

Now, a grass-fed cow accumulates beta-carotene in fat. It's why cow's cheese goes a golden hue, and cheese from sheep and goats doesn't, because those animals don't accumulate beta-carotene. This golden colour is what can make the fat in the meat from a grass-fed cow more yellow – considered for a while by some as a bad sign, though good farmers always knew. Beta-carotene is the reason butter is yellow, or used to be yellow. You may notice butter becoming paler and paler every year as cows get put indoors more and more, and eat less and less fresh greens.

All that golden colour is a neat little visual cue; an indicator of grass feeding. Imagine if you could see all the other phytochemicals and micronutrients that result from a varied diet of forage, herbs and grasses in your yoghurt, compared to milk from cows locked in a shed? Imagine if you could see 20 times more micronutrients in your butter. And imagine what the difference would be for your body.

Buttergate

In 2021 there was a big dairy controversy in Montreal, Canada. Milk wasn't frothing properly for their café au lait. Cheese had a weird texture. And, *quelle horreur*, the butter was too hard. The French-speaking, butter-loving Québécois were aghast. Was it the cold weather? Had the producers changed the way their butter – a product the locals consider a natural food – was made? *Le Journal de Montréal* ran an exposé, telling the public something the dairy industry already knew: that palm oil products were being used in some dairies.[20] It was, *Le Journal* declared, a 'broken moral contract' to add such a thing to a cow's feed and alter the very nature of the product she produced.

There was consternation galore. It must be the palm oil, cried the masses, outraged that someone was importing a controversial product from the other side of the world to feed to Canadian cows.

Why palm oil? And how could it possibly be responsible for a change in the texture of butter and milk?

Palm oil is hard at room temperature. It has a lot of palmitic acid in it – a fatty acid that occurs in decent amounts in coconut oil, too, and also exists naturally in butter. In fact, it's the main fatty acid in butter (albeit in far lesser amounts than in coconut and palm oil), and, being saturated, it's solid at room temperature.

The palm oil industry is massive, blamed for wholesale rainforest destruction and ruining the habitat of orangutans and other wildlife. It takes up 77% of the agricultural land in Malaysia, for instance. Palm oil is cheap to make, and way more efficient per hectare than other oils in terms of yield. The world produces more palm oil than any other kind (second is soybean oil, and a long, long way in third is canola/rapeseed). Palm oil is used in everything from biscuits to cakes, from pasta sauces to curry pastes. It's in half the food products on the supermarket shelf. And, as we saw with soybean and oatmeal waste, that leaves a problem: 75 million tonnes of palm oil produces 190 million tonnes of effluent that is left behind in the countries that produce the oil. And there's also 400 million tonnes of 'palm oil expeller' or 'palm kernel expeller' – the solid part of the palm kernel left after the oil is extracted. Much is mulched to give back to soil. But it still contains some fat and protein, making it a cheap feed supplement, especially for grain-fed dairy cows in sheds. It's tempting to upscale waste into something valuable, like milk. Problem is, that palm oil expeller still has lots of oil in it – up to 20%.[21]

Could this palm oil really be the reason for unfrothy cappuccinos, weird cheese and hard butter in Canada?

Into this breach stepped science, in the form of Alejandro Marangoni, a professor from the University of Guelph. He gathered up 51 different Canadian butters. He melted them, analysed them.

He inserted a diamond-shaped tool into each pat to test its resilience. And lo and behold, he found that the firmness of the butter and the amount of palmitic acid in each showed a remarkably strong correlation. So, no, it wasn't just everybody in Montreal feeling weaker post Covid, or some new manufacturing technique. *Sacre bleu*, the butter *was* harder, and the diet of the cows *was* to blame as dairy farmers turned to palm oil and its expeller as a source of feed.

Just a few short years prior, the palmitic acid content of Canadian butter had been 26%. In five years, it had risen across the board, in some butters up to 40%.

Now, I'm a big fan of not wasting stuff, but only feeding it to animals if it is likely to be part of their natural diet, or the consequences are known. And one consequence of palm oil is hard butter in Montreal – and elsewhere. I've seen palm oil and expeller waste being used in dairy rations in a few places around the world, including the UK, Australia and New Zealand; even those cows on grass all day may be fed palm oil in the dairy when they're being milked.

Palm oil doesn't just affect the texture of milk. It's probably not very good for the cow's biology, either.[22] We are yet to do the research on what it does to a cow's health, but palm oil in expeller or in its refined state does contain more than a dairy cow's tolerable levels of iron, phosphorus and magnesium, and puts at least nine nutrients way out of whack, including increased intake of potassium, boron and aluminium.

Should you trust your butter? Well, in the US, they feed chicken shed waste to dairy cows. And palm oil. And they feed cows tallow – rendered beef fat,[23] which seems a bit cannibalistic to me. Last time I looked at my dairy herd, nobody was eyeing off anybody else's fat reserves, so, you know …

Chicken feather meal and tallow are banned as feed for ruminants in Australia, Europe and the UK, thankfully. But palm kernel expeller – and palm oil more generally – is used everywhere.

Now, palmitic acid isn't one of those fatty acids that has miracle properties. It's been associated with all kinds of problems, from tumour progression to blood and psychiatric disorders. About

20–30% of the fat in our bodies is palmitic acid, so its role in disease is a bit hard to decipher. It seems if we get the amounts of the fats we eat all out of whack – including what we eat it with, and its ratio to other fats – then it can cause health issues. Just another reminder to balance your fats. And perhaps a reminder to dairies to not put palm oil, or beef tallow, in their cows' feed.

Is it palmitic acid that plays into the dated idea that saturated fat will give you heart disease? Perhaps. I know I still hesitate before slathering a greedy amount of butter on my crumpets. Should I be more careful with the lashings of clotted cream on my porridge? Maybe, if my biology says so. Is cheese a heart attack on a plate? The experts are clear on this.

I'll defer to a review on saturated fats and health published by the *Journal of the American College of Cardiology* in 2020:

> **Whole-fat dairy, unprocessed meat, and dark chocolate are saturated fatty acid-rich foods with a complex matrix that are *not* [my emphasis] associated with increased risk of cardiovascular disease. The totality of available evidence does not support further limiting the intake of such foods.**[24]

Though perhaps they weren't expecting palm oil in their butter.

Breastmilk Without the Breast

I've never milked a reindeer. Not many have, because the amount of milk you get is barely worth the effort. You can only entice a single litre, or less, from a reindeer in a day. Not much more than a lactating human.

Why do it then? Reindeer milk is dense with nutrients. It's about 22% fat – nearly eight times that of human milk. And it's about 8% protein – roughly 10 times more than was in your mother's milk. Reindeer milk needs to be rich, so their calves can mature quickly in the short northern summer. And people milked reindeer because above the Arctic Circle, everything you can find to eat is valuable, and no other dairy animals can survive in the climate. Reindeer milk was watered down and drunk fresh by children. It was made into curds in another reindeer's stomach and set, ready to sustain the family over winter and spring. It was soured and mixed with aromatic herbs. Reindeer milk was turned into cheese and butter, the yield far greater than cow's milk per litre, because of its high protein and fat count. And the buttermilk and whey that remained were boiled down and eaten as soup. Nothing from the reindeer was wasted by the Laplanders of northern Europe.

Reindeer milk isn't much of a thing these days. Some indigenous groups still use it for their own purposes, but it's not a commercially viable product. The only reindeer milk I found for sale was a powder in New Zealand, which works out to be about A$28 (or US$18.50) per litre when rehydrated. It's certainly not cheap.

Camel milk costs about $17 a litre fresh in Australia, and about $300 a kilo dried. Moose cheese in Russia is worth about US$1000 per kilo, apparently, and you never see moose milk for sale fresh. There's a reason we settled on cows, goats and sheep as our main dairy suppliers. They produce lots of milk for relatively low effort.

As we saw earlier, rare milk is worth a lot of money. None as much as breastmilk, though. While a few are giving away their excess, others are selling it for up to $500 a litre. We know it's good, and we know it's in relatively short supply.

Despite the formula industry being worth an estimated US$70 billion each year, we still haven't managed to replicate even just one important component in breastmilk, the non-digestible milk sugars – human milk oligosaccharides – in the lab. We could well put a human on Mars in my lifetime, but milk's complex code is still waiting to be cracked.

But there might be a way.

Milk contains lots of cells. Like, LOTS. Researchers reckon there's 23,650 cells in every millilitre of a healthy woman's breastmilk. (And that figure is actually way less than the 146,000 cells per milli-litre found in colostrum.)[1]

These cells aren't pus, unlike what groups like PETA would have you believe are in cow's milk. To the contrary, the cells include all those wonderful leukocytes we met earlier, along with all those vitally beneficial bacteria to inoculate the gut, and the immuno-globulins and other epithelial cells from the mother. There's also another type of cell that could prove valuable in the feeding of premature human babies: stem cells from the mother.

For a while we didn't know stem cells were in milk. Stem cells are unlike other cells in the body. They have the ability to differentiate into all kinds of other tissues in the body – to become hair or skin or muscle or the lining of the gut. Or breast tissue. In fact, scientists have used stem cells from teeth to produce mammary gland tissue.[2]

But we also know breastmilk contains actual breastmilk cells, too – and we now have the ability to culture them up in labs.

Using these breastmilk cells, scientists are breeding cultures that have the potential to create a more complex milk than yeasts ever could. In research that mirrors some of the technology used to put fake meat on the market, mammary gland cultures are being used in tests to create a form of breastmilk completely unlike formula or something you'd get from plants. This is the closest we've got, so far, to re-creating real milk.

Originally, this kind of research was designed to help in organ transplants, using a person's own cells to re-create the organ in need. It then hit the news when Dutch researcher Mark Post showed the world the first cell-cultured burger (which at that stage cost US$325,000), and later when an Australian 'cultivated meat' company used cells from the long-extinct woolly mammoth to create the world's first (at least in a long time) mammoth meatballs in 2023.

Leila Strickland is a woman making breastmilk. Nothing new there, I suppose. The difference is she is making a *form* of breastmilk outside the body, at least in part. She helped set up Biomilq in North Carolina, using a lab-based cell-culturing system to harness milk from laboratory-formed breast tissue.

The way milk is produced in the body, which we looked at in the chapter How Real Milk is Made, is a complex process that is remarkably consistent across species. Breast tissue lies dormant until triggered to reproduce by a mother's hormones during pregnancy, readying her body for the birth. Spaces are formed in the glands for the milk to be stored in. Then other hormones initiate the production of colostrum, then milk, within the cells. In our bodies, and in the bodies of other mammals, lactocyte cells do that, using blood as their nutrient supply. This is what Biomilq are trying to mimic. They're breeding breastmilk cells in the lab until they're in large numbers, then creating a bioreactor – essentially a big fermenting tank – that is lined in a way that the cells can coat one side of a porous film. Nutrients are fed in one side, and a form of 'milk' comes out the other side. To help get this far, they've raised millions in funding, including US$3.5 million from Bill Gates.

Strickland has been working on this for over a decade. According to the American cable news network CNBC, in time she hopes to be able to synthesise human milk's 'complete profile of macronutrients', while at the same time meeting the 'functional definition of milk from a composition standpoint'.[3] In other words, get the fat, protein and sugars in proportions and in composition pretty close to human breastmilk, and ideally also have a few of the other factors that make milk so amazing.

This technology is years from being released to the public in products we can buy, particularly in infant formula, because it has to be proved safe. The other problem is overcoming fear or distrust in this space, something lab meat has struggled to do.

Biomilq isn't the only company looking at producing milk from animal cells. And it's not just human cells that researchers are using – it's also those from our ruminant friends.

In Singapore, there's a mob called Turtle Tree, headed by Fengru Lin, hoping to do similar things. Like Biomilq, they're sounding more humble than most plant-based milks, for a start recognising that there are over 2000 different components in real milk. Lin was inspired to get into cell milk after learning to make cheese in the US, only to head back to Singapore to realise there's no dairy on the island state, so every drop has to be imported.

For the moment, Turtle Tree is building up its research facility in Sacramento, California (where they actually have access to cows, if they wanted to use them). They're also focusing on using GMO yeasts to produce lactoferrin, the incredible biochemical we met earlier. In the background, cell culture using animal tissues is where they'd like to be at. They've already succeeded in creating experimental milk products using cells from cows, goats, sheep and camels.[4] Apparently Coca-Cola Israel is a fan, stumping up $2 million in seed funding.[5]

Coca-Cola Israel has also invested $2 million in their local animal cell-based milk start-up, Wilk. Founded in 2020, Wilk is keen on moving cows out of dairies and milk into labs. Recognising milk's complexity, they are also working in the breastmilk space[6] (though they don't suggest they'll be replacing mothers).

'Milk can never be replaced', Wilk acknowledges on its website, but it seems they believe they can *replicate* it in a lab, which is ambitious. They're moving fast, already producing a yoghurt that contains cell-cultured milk fats, and was taste-tested in late 2022. I contacted them to ask about the ingredients in the yoghurt, including the fat, but unfortunately they did not respond.

For much of this book I've talked about nutrients. But humans aren't designed to eat nutrients. We're designed to eat food. Several million years of evolutionary fine-tuning has made it thus. The magic fats – the CLAs we met earlier, for instance, which can switch on cancer-fighting cells – can be damaging if eaten in excess. Alpha-lactalbumin only becomes the cancer-fighting HAMLET when it's in combination with the good fat, oleic acid, which milk also contains lots of. Everything in food has an effect on our bodies, in combination with *other* things in food. If we want to replace real milk with plant milk, or cell-based milk, we need to put back what is in milk in comparable amounts.

To replicate the 2000-plus micro and macronutrients in real milk using 'nature inspired technology' will take some doing. If you use yeasts, you'd need 2000 separate bioreactors, churning through electricity, using embedded energy in the tanks, needing a feed-stock of refined sugars, some kind of nitrogen and perhaps oils, and creating masses of dead yeast cells in the process. If you use animal cell cultures it's even more exacting, requiring medical-grade facilities, and with as-yet unknown implications, both on us, and on the environment that will be mined to feed those cells.

Cell cultures are also working in competition to – but sometimes also in tandem with – microbial fermentation. If we can synthesise them, then those 2000 ingredients would need to be extracted safely, blended in exact proportions, the 'milk' kept from contaminants at every step along the way. And that's basing the new fake milks on what we know about real milk right now; there's still lots we don't know about it.

Is something so complex even possible to replicate? We know we've never come close with anything else humans have manufactured, so it's a big ask. If the complications and expense of lab-based meat is anything to go by – with the need for new factories estimated to cost roughly US$450 billion just to replace current beef consumption in the US alone[7] – the production of these complex milks is unlikely to be able to compete with real cow's milk at the checkout anytime soon.[8]

Good Meat, the first company to produce chicken protein from chicken cell cultures to sell to the public in late 2020, still only produced enough to feed 20 people once a week from one butcher in Singapore in mid-2023.[9] Even then, 63% of fake chicken isn't even chicken cells: it's filler and gums and fats. A burger using cell-based chicken meat costs US$100, even using just 10% meat cells. If this is the case with meat, which was ahead in this space, it's hard to see cell-cultured milk being much different. It may also have supply issues and cost blowouts just as lab meat does.

While many in this space frame their concerns about real milk and meat as environmental sustainability and animal welfare problems, much just looks like profiteering. As Joshua March, co-founder of SciFi Foods, a company intimately involved in the cell-culture meat space, puts it, 'Ultimately, everyone's trying to create a start-up that survives and can raise funding and has its own IP [intellectual property] and trade secrets.'[10] There's little sharing of knowledge, little collaboration. Mostly it's driven by competition.

The problem for humanity with the current breed of plant milks is that they offer just a fraction of milk's nutrient density, masquerading as 'wholesome' and 'natural' when they're the products of the industrial food system – items at the whimsy of food technologists that represent the worst greenwashing of ultra-processed food. If we move to cell-based milks we also come up against logistical, practical, cost and health implications, and an industry riven by the need to own the technology.

This needs to be looked at in real terms, given the fact that, if current dietary trends continue, we'll need to produce the same amount of protein in the next 30 years as we have in the last 2000 years.[11]

We already have a natural bioreactor that ferments sugars that we can't eat and turns them into a mind-numbingly complex liquid in mammary cells. It's called a ruminant.

That liquid the ruminant produces may not be fully digestible by all. It isn't necessary for a healthy diet. Milk may have an environmental cost that is too much for humanity to bear, especially when we, as a species, drink more dairy globally, and if we cut down more trees to nourish a growing population.

But replacing it with fake milks with hollowed-out nutritional profiles that do dubious good, and consume resources that could be put to better use doing other things? Well, to chase that dream seems to be the quest of those in money markets and tech disruption, not those who actually work feeding the world. So far with substitute milks, in terms of the global food supply, it's all talk, and not much to eat or nourish us. Will this change? Economics may well play a part.

According CNBC, 'Biotech funding reached a record high of $77 billion in 2021 ... but it then dipped 38.6% between 2021 and 2022.' The shallow, fickle nature of this industry was apparent when investment in it crashed after the failure of Silicon Valley Bank.[12] You can want something to be true, like many did after the much-vaunted RethinkX report from 2019 predicted that 50% of American cows would be gone by 2030, and 90% by 2035. Numbers of US cattle *have* declined, by about 7% in three years, but analysts put that down to dry times and lack of feed, as well as shortage of labour in meat-packing plants. It's not down to the proliferation of meat substitutes, no matter what some might have you think.

Will the same be true of fake milks? They're certainly quicker to find market share, though milk's falling consumption is also due to higher sales of soft drinks, juices, energy drinks, and the original con, bottled water. As we've seen, when we look at recent falls in dairy consumption, plant milk sales only account for about a quarter of the drop in milk sales. They aren't filling the gap. Sugary drinks are. Perhaps that's why companies not necessarily with nutrient density and public health as part of their mission statements, including PepsiCo and Coca-Cola, are investing in the new market.

I Can't Believe It's Not Better

You can tell I'm sceptical of the hype. I look out my window at a region that for the last 200 years has been dominated by apples and see grazing land – land that would become a death trap in fire season if not managed somehow. The obituary written of ruminant farming seems premature from the regions of the world, if not from the cities where most commentators sit, toiling with their pens and their mouths, not with their backs or their hands. Ruminants do more than just make meat and milk and emit methane. They perform landscape functions. Yes, they can be damaging, but they can be healing too.

That's not to say fake milk won't find a place. It already has. Out of six milk fridges I spied in a supermarket last week, one was mostly plant milk. That amount of shelf space makes sense. It's great to see dairy-free options for those who are intolerant, allergic, or just avoiding dairy for whatever reason they choose. But, milk they ain't.

What lessons are there in history? If we look at the great impostor in our midst, margarine, the original plant-based fake dairy, it had half the fridge space to itself in the supermarket I visited recently. Despite a dubious reputation for its role in heart disease thanks to trans fats, despite the fact that most thinking people don't rate the flavour (they even try to put fake butter flavour in it), despite the ultra-processed nature of margarine, it still has a pretty healthy market share. It trades, globally, at US$33 billion compared to butter's US$46 billion, though since 2004 people in the US have been favouring butter again. Much of the trade in margarine is hidden, of course. It lurks in the ultra-processed food industry – in your bought biscuits, pastries, cakes and ready-made meals.

In the UK, margarine sales have nearly halved in the decade since 2011, while butter sales have risen.[13] The tide has turned in Australia, too. In 1992, three-quarters of sales of spreads went to margarine, and only a quarter to butter. But this has swung back.[14] We were duped, but not for long. By 2018, nearly 70% of Australians buying a spread bought butter or a butter blend, and only 30% bought margarine. It's not hard to imagine the same will happen with other fake dairy. It will find its place, probably like margarine in the processed food industry. Someone will produce it in a factory and make some money off it. But it won't have improved us gastronomically, and I worry that it won't have improved us nutritionally. Or improved the planet much. I lived through the margarine years, and I'm still scarred.

Margarine's lower price than butter obviously helps its sales. But what about fresh milk? Is its death notice premature? Well, 98% of British households bought milk in the last year for which figures were available, up to 2023. They bought milk 62 times that year, a drop from 64 times in the year prior, so hardly a drop off a cliff face despite the rise of plant milks, especially when you consider fresh milk's price rose 27% in that same period.[15]

You may hear about the death of dairy, but is it just wishful thinking on behalf of some who have more media clout than they should?

While cell-based and yeast-created milk is fraught technically and financially, it also has the possibility to do incredible good. We are now learning how to produce the parts of milk that are currently used in ultra-processed food, without having to resort to industrial dairies. This kind of ingredient processing may be useful for 3D printed dairy products and the like (yes, there are such things – there's even a prototype Cadbury Dairy Milk chocolate printer[16]). Perhaps we can also identify some big nutritional gaps in our processed food that can be filled with simple proteins, fats or even lactoferrin from factories. But, thanks to costs and logistics, analysts say these 'milks' aren't likely to be the answer to milk anytime soon;[17] they're just another answer to ultra-processed food.

The great hope for cell-based milk in particular isn't with more processed food for adults, but to nourish preterm and at-risk babies better. Perhaps cell-based milk from yeasts and animal cells can fill the gap. Remember Brazil? Despite having nearly a third of all breastmilk banks in the world, the country still finds 40% of their preterm and low-birthweight newborns are undernourished. We've always had difficulties feeding babies. Anything we can do to make that better is a good thing.

Will lactoferrin and human milk fats and proteins from cell cultures and labs make it into baby formula more generally? Probably. Is that a good thing? Yes, perhaps, because it makes formula more complete as a food, even though it can never be breastmilk's equal.

But it's not a good thing if it encourages less nursing at the breast. In the 20th century, the rate of breastfeeding around the world was

90%, but in the first few years of the 21st century, it has already dropped to dangerously low levels globally, to approximately 42%.[18] We know breastmilk is best. Even Biomilq talks up the positives not just of breastmilk, but of breastfeeding, for all the biological, immunological, sociological and psychological benefits that putting your baby on your boob provides above putting them on the bottle.

Milk isn't just nutrients. It's nourishment. When we're infants it has all the things, thousands of chemical components, that are doing us good. Milk provides two-way communication between our bodies and our mums. It inoculates us, protects us, nurtures us, alters the expression of our genes, and connects us to our environment.

The milks we drink as adults, and the dairy we consume, can also play a role in our brains, our guts, our nerves, our moods. We don't need milk after weaning to live healthy, vibrant lives. But if you don't use dairy, to get the complex interplay of things that milk provides, you'd want to make sure you're eating a wide variety of many things, all through life. Replacing that with something less vital and made in a factory is risky when our diets have already simplified, our connection to the land that grew the food is more tenuous, and many consider cooking a badge of servitude best left to others.

Is real milk good, or bad? Is cell milk good or bad? Is plant milk good or bad? They all have their flaws. They all have consequences – on us, on our communities, on our planet. Balancing all these considerations is a decision for each of us to make, which will vary depending on our physiology, our genetic makeup, our geography, our moral compass.

Just beware the marketing spin of those with large budgets, from both sides of the dairy debate. Like all things to do with milk, if it's simplified into a soundbite, it's probably wrong.

The New Moonshine: Raw Milk

'You got any?' The question came breathlessly. A little shyly.

The surreptitious nature of the query reminded me of the 'Jason?' I used to hear whispered behind me on Smith Street in Collingwood a few decades ago. *Jason?* I wondered what the man behind me in a tracksuit was trying to say. Turns out he was asking if I was 'Chasing?' But I was too naïve to know a drug dealer when one followed me. Too busy looking for something to eat to be 'chasing the dragon', looking for smack.

'You got any?' I'm left wondering if people think we're growing marijuana in our market garden, a crop our region was once quite well known for. A crop that certainly makes more money than vegetables, but one we are yet to propagate.

But no. It's not dope they wanted. It's milk they were after.

A local dairy had been busted selling raw cow's milk as 'pet milk' and the authorities weren't happy. In Australia it's illegal to sell raw milk. Near us, a one-time commercial dairy – a place that understands cow health, cleanliness and tests their animals regularly – had been told to stop selling milk for pet food or cosmetic purposes (bath milk) to locals. And suddenly I realised just how many people are buying raw milk, not just to feed their pets or bathe in, like Cleopatra apparently did, but to drink.

It's become a black market, and milk is the new moonshine.

Because we milk a cow, we're on a clandestine list of potential suppliers. We don't sell our milk, just use it for our own consumption, but I'm still witness to dozens of people who are mad keen on getting raw – unpasteurised – milk. I watch as they head off from us to smaller, perhaps less hygienic sources, to get their fix. They think it will cure their eczema. Or their depression. Some believe it may help with their cancer, or their child's learning difficulties. Others are sure it will fix their microbiome and heal whatever it is that ails them this month.

They may have a point. We know that people who drink raw milk have less asthma and other allergic responses, as well as fewer respiratory tract infections such as colds and flu.[1] Research from Germany is exploring how minimally processed milk (including minimally pasteurised) can help everybody avoid allergies, and asthma in particular, and the results are promising.[2]

Whatever the reason, a lot of people really, really want raw milk in their lives. There's a huge underground market in cow's milk, goat's milk and breastmilk, so long as it's not heat treated. I know a young barista who lives in a share house and meets a farmer in a McDonald's carpark once a fortnight to buy raw milk because it's all the rage in his community. Raw milk is also huge on TikTok, apparently, which is always a cause for concern.[3]

Now, I have an odd view on this. I drink raw milk. I think farmers should be able to sell raw milk, so long as they run a very tight ship in terms of cow health and hygiene. I think customers should legally be able to buy this milk, but only at the farm that produces it. And what I really, really think is that all those customers shouldn't buy it.

Kurt Timmermeister once ran cafés of ever-increasing size in Seattle, in Washington State. Until, one day, the American chef bought a farm on nearby Vashon Island and started milking cows. In his book *Growing a Farmer*, Timmermeister describes how hard it is to navigate the issue of dairy. He was selling raw milk, illegally at first, to people who discovered he milked cows. After a food scare in a nearby state, he decided to run a properly licensed dairy and make cheese and butter. In the process, he learned a lot about how to handle milk, to ensure it's safe. In Washington State (and 11 other American states, apparently), you can actually sell unpasteurised milk legally, so Timmermeister became a licensed raw milk vendor.

But not long afterwards, he stopped. Not because he didn't produce it safely, but because he didn't trust his customers.

Raw milk can be a hot topic in food, farming and nutrition circles. I've got a mate, Nick Haddow, who makes raw milk cheese. He's made cheese in France and Italy and Ireland. He is a big advocate for changing the rules in Australia to allow us to make raw milk cheese here – cheese of the style we seem happy to import, but is impossible to make locally. At least legally. And for a long while, Nick used to wear a T-shirt that said, 'Raw milk doesn't kill people. People kill people.'

Nick's public statement echoes Timmermeister. Milk isn't designed to be pathogenic, in other words dangerous, to a calf. If the mother is healthy, and the milk is fresh, then it's probably going to be – by definition – safe. Either to the calf, or to us as adults. It is, again by definition, designed to be nourishing.

What happens after the milk comes from a cow, however, is another thing altogether.

While some people are very happy to believe that raw milk is a superfood, incredibly nourishing, full of beneficial bacteria and potentially immune-boosting proteins, they can also forget to put it in the fridge. Or neglect to drink it soon after it came from the cow. I once looked up a raw 'bath' milk available in Western Australia, where it had a two-week use-by date and could be delivered to your door by an online retailer. (They did helpfully suggest they could leave the package in a shady spot if you weren't home. Perth, for those who don't know Australia, regularly endures days over 30°C/85°F in summer.)

Now, raw milk is full of bioactive compounds and living biota. Some good, others maybe not so good if they suddenly grow in large numbers. Mostly, it helps to think of milk as not just nourishing to us, but also nourishing to all the things that would potentially kill us – including dangerous pathogens such as *E. coli*, campylo-bacter, cryptosporidium, salmonella, listeria. In some places there's tuberculosis and brucellosis, too. After all, it was raw milk that killed a lot of people, particularly infants, in the 1800s and early 1900s.

In theory, we can learn to look after raw milk, in the same way we have learned to handle potentially poisonous foods such as raw seafood, chicken and eggs. We're trusted to be able to not kill ourselves with those, which are inherently more dangerous than milk in most cases. But it means re-thinking how we look after and feed the cows, how we milk the animals, and how that raw milk is distributed.

Let's look at pasteurisation and what it does for milk – and then how or whether raw milk could be legalised.

We can thank wine for pasteurisation. Those wine-loving French folks of the 1800s had problems with some of their Bordeaux and Champagne and Bourgogne. Sometimes, things went off.

They enlisted science in the form of Louis Pasteur, a chemist who found that it was living things, the yeasts and bacteria, in wine that made it go off. And he discovered a way to kill them, at least enough to prevent wine turning sour. One beneficiary of this research was milk.

The technique of killing potentially dangerous and food-spoiling microbes in milk and other products still carries Pasteur's name. Pasteurisation is a method of using heat in combination with time to reduce the risk of bad bugs, and it's been a blessing that has allowed dairy to remain not only popular, but safe.

Usually, milk is pasteurised by being heated to about 72°C (162°F) for 15 seconds. It is done, predominantly, using a heat exchanger, where the milk runs through tubes, taking 15 seconds to reach the end of the pipes. As it flows, the heat from a boiler is drawn into the milk to get it up to temperature. After the regulation 15 seconds, it is then passed through cold pipes, so it's cooled again within seconds. It's called HTST, which stands for high temperature, short time.

There's a sliding scale of death, however, so you can pasteurise at a lower temperature, say 63°C (145°F), but you have to heat it for way longer, for 30 minutes. This means you can't just run it through pipes, as you'd need a very, very long pipe for milk to be pumped through for half an hour. This kind of pasteurisation is called 'batch' pasteurisation. You heat a whole batch of milk, from 20 litres (4 gallons) to 2000 litres (440 gallons) or more, to the right temperature, then leave it, stirring constantly, while it pasteurises. When done, it's pumped out and more milk can come in.

For practical reasons, heating the milk to 72°C (162°F) for 15 seconds through a heat exchanger is usually favoured.

Why am I telling you this? Because we know that heat affects more than just the bacteria in food. It can affect the structure of proteins and fats, too.

Both methods of pasteurisation work to make milk less likely to carry the wrong things. But extra heat also has other unintended immunological consequences. For instance, while pasteurisation kills the bad bugs, it also kills all the good bacteria, too – leaving the milk a blank canvas for the first thing that might colonise it, good or bad. With nothing to outcompete them, bad pathogens can flourish. Luckily, we also have refrigeration, which slows down the growth and breeding of all bacteria.

With pasteurisation, we also lose flavour. The reason you see 'cold-pressed' oils, like olive oil, sold at a premium is that complex aromatic esters are quite volatile. They blow off as the oil is heated.

That's the reason cold-extracted honey is the gold standard, because heating honey much above the temperature of the hive (about 30°C/85°F) will render the honey less flavoursome.

The same is true with milk. The more we heat it, the more nuances we lose. The flavour of the pastures. The subtle scent of the cow. The flavour of the season. And we add unwanted flavours. Bitter almond. Off, fishy characters. Things that remind us of old hessian bags and the like. It's minor, and you may be used to those flavours, but they're not the same as those in raw milk.

That's why my mate Nick is adamant that we should be making cheese, safely and where it makes sense, with raw milk. When milk becomes cheese, these aromatics are multiplied 10-fold, because it takes about 10 litres of milk to get 1 kg of cheese. This condensed version of milk has trapped all the subtleties and charisma of the original product. It traps the season, the nature of the single herd, or the breed of cow; the taste of place. *Terroir* it's called in French – the way you can perceive the land, the soil, the climate, the geography, the variety of the thing used (grape variety, breed of goat) on the palate. It might sound fanciful if you haven't milked animals and tasted the result of their lactation, but there's a clear difference in milks, one that is lost once blended and boiled.

And when I say boiled, most milk is only heated to 72°C (162°F), which takes some of the romance away from the flavour – but some is heated to 140°C (284°F), to make UHT ('ultra-high temperature') shelf-stable milk, which caramelises sugars and deadens any herbaceous notes.

Some people like the burnt sugar flavour of UHT, but I think it's less interesting, and also has more unattractive bitter notes. The science backs this up, with various aromatic acid compounds being greatly reduced the hotter that milk is heated. Other things, called ketones, esters and hydrocarbons, start to stick together more. Essentially, the more you heat milk, the more you lose in terms of complexity and original flavour.

Milk that is heated at lower temperatures for a longer time – so-called LTLT pasteurisation – is likely to carry more flavour. The fat will glob together less, too, meaning it's easier to shake the cream back in if you buy unhomogenised milk.

The magic number seems to be 65°C (149°F). Anything higher than that and you start to get more denatured proteins, more sulphurous compounds (cabbagey smells), more furfural (baked aromas) and the potential for stale and rancid characteristics.[4] For many, that flavour is what they are used to in milk. The hotter milk is heated, the more the effect.

Of course, heat also changes the nutritive nature of the milk, and the way it is absorbed into our bodies. The fats are digested slower, and less overall,[5] while the protein in pasteurised milk clumps together less, so that after about four hours, it's 70% digested compared to only 20% for raw milk.[6] What this means nutritionally is still a matter of scientific contention, but it does impact blood sugar, and our sense of fullness – and it's likely that we reduce the protein bioavailability when we heat milk.[7]

There's a new way of pasteurising milk that keeps some of the flavour compounds intact. It's cold pasteurising, where the milk is put under high pressure – high enough that it splits apart the bacterial cells, killing the pathogens as heat would do. Does it change the milk? Of course it does. The fats clump together differently, some proteins are denatured, it looks more golden, but the sweet character of raw milk emerges pretty much unscathed. Is it as bioactive? Well, no. Is it safer than unpasteurised milk? Well, so long as you don't contaminate it later, then yes.

High-pressure pasteurisation is looking like it'll be really useful for other reasons, too. Research has shown that the method milk banks use to pasteurise human breastmilk, called Holder pasteurisation, where milk is heated to 62.5°C (144.5°F) for 30 minutes, kills 99.999% of bacteria – but it also lessens the effectiveness of lipase, lactoferrin and immunoglobulins A and G. Premature babies fed pasteurised breastmilk could only absorb 60–70% of the milk fats compared to unpasteurised milk, too. Using high-pressure pasteurisation without the heat is as good as Holder for killing bugs, and yet does less damage to immunoglobulins, lipase, lactoferrin and more.[8]

Again, this is very recent research, but we should know its effect on fat absorption rates soon.

If milk is generally safe when it comes from a healthy cow and is well handled, what's the problem? Shouldn't we all be able to buy better tasting, potentially more bioactive unpasteurised milk? Well, like all of this milky stuff, it's complicated.

Nothing comes without risk, of course. In the US, where raw milk sales are possible in some states, they collect really good data. It's suggested there's 3.2 times as many outbreaks of illness from dairy products where raw milk sales are allowed than where they're not allowed. And in states where they allow raw milk sales, you're 3.6 times more likely to get ill if you buy the milk from a shop, rather than at the farm. So, regulation seems to equal less illness.

Some estimates put it that raw milk products are 150 times more likely to make you sick than pasteurised milk products. That sounds terrible – but because almost no-one gets sick from pasteurised dairy products, the numbers are actually quite low. Let's break it down. Each year between 1993 and 2006 saw an average 10 outbreaks of all dairy-related illness in the US – with 60% of those due to unpasteurised milk. An average of 157 people a year (1571 people in total) became sick from raw milk or dairy.[9]

In the US in that period, about 80% of American adults consumed dairy every day,[10] meaning 240 million adults[11] consumed dairy 87.6 billion times a year, assuming they ate or drank dairy only once a day.

Now, I'm not going to pretend raw dairy is safe. Nothing is. But 157 illnesses arising from the 87 billion times people ate dairy is pretty small – 0.0000017% of a chance, actually. That's with 4.4% of the adult population (about 11.5 million people) consuming raw milk in the US at least once a year.[12]

Dairy is very, very safe. So, a tiny risk for all dairy, multiplied by 150 times the *extra* risk for raw dairy, still means a tiny risk.

Let's put this in context. Between 1998 and 2018 in the US, there were 202 outbreaks of food poisoning from raw milk compared to only nine from pasteurised milk. So, on average, about 10 outbreaks a year compared to less than one. But get this: over those same decades, there were 21,919 outbreaks of food-borne poisoning. Close to 22,000 outbreaks of food-borne illnesses, and only 202 from raw milk – with raw milk accounting for 0.9% of all food-borne outbreaks.[13]

And, when you look at how many people actually became ill (the outbreaks represent a product or foodstuff, the illnesses are how many people got sick), unpasteurised milk made 2645 people sick, pasteurised fresh milk made 2133 people sick, and food more generally made a whopping 423,595 people sick. Of course, way more people drink pasteurised milk than raw milk, so you have to factor that into those figures as well.

In case you were wondering (I was), the most common sources of food-borne illnesses are fruit and vegetables.[14] According to a 2018 paper in the journal *Epidemiology & Infection*, which looked at food-borne disease outbreaks in the US:

> **During 1998–2013, there were 972 raw produce [fruit and vegetable] outbreaks reported resulting in 34,674 outbreak-associated illnesses, 2315 hospitalisations, and 72 deaths.**[15]

While the milk-specific study went over a slightly longer time frame, fruit and vegetables were approximately six times more likely to cause a food poisoning outbreak, and nearly 20 times more people actually got sick.

Everything comes with risk. Often I think the most dangerous thing about raw milk is the drive to pick it up from the shop. Or getting into an argument with someone about it.

Q. What do a shooting star and milk have in common?
A. They're both past-your-eyes before you see them.

Pasteurisation is a brilliant tool that lets us move milk around more. To store it for longer. To let you off the hook if you leave your milk out of the fridge for an hour after making a pot of tea, or if you don't drink your milk within three days. It allows those of us who run dairies to be slacker with the milking (no cleaning of the udders). It allows your customers to be loose with the shopping (leaving the groceries in the car while doing something else for an hour), and cavalier with the hygiene at home. That suits most people most of the time.

It does mean you miss out on flavour. You probably miss out on some of those bioactive compounds in milk – the EVs, the peptides. And you don't get the full benefit of the fat.

Humans can produce raw milk in a manner that is just as safe – if not safer – than the oysters and chickens we are happy to let people buy, which cause way more illness than milk. But that might mean a whole new way of looking at things, from the cow, to their feed, to the way milk is distributed. It certainly would need a big education campaign.

The big players don't want it to happen, because it's in their interests to sell lots of milk from lots of big dairies to lots of people who live a long way from the farm. But the reality is that some people are already risking their lives looking for the new moonshine. And when you drill down into some of those food-borne illnesses from raw milk, it's from people making cheese in their bathtubs. Seriously.[16] So-called bathtub cheese made from raw milk made 2000 people ill in Utah in 2011 when they came down with salmonella.[17]

If we push things into the realm of the unregulated, into the black market, then we have no say and no idea whether the person selling you stuff out of the back of a truck or making cheese in their barn is across milk hygiene, herd health, sanitising equipment, and the speed at which we need to chill milk after it comes from the cow.

We also can't help those who want to produce dairy safely to avoid the pitfalls. What we're doing now isn't working because people actively break the law.

Is pasteurisation needed? I think so. Should raw milk be legal for sale directly at the dairy farm? Yes. Should you buy it? Only if you really understand the risks. Do we need to rethink our incredible fear of raw milk? Definitely, yes.

Regulating and selling raw milk today wouldn't look like it did 100 years ago. Refrigeration took off at the same time as pasteurisation, and it's just as vital for making milk safe. Refrigerating milk from the cow to the consumer made a huge difference. So did the understanding of basic hygiene. We wouldn't get the same result today as we did in the 1800s if we abandoned pasteurisation, but we'd still probably get way more illness from dairy than we do today.

Is that a risk worth bearing? As we saw above, pasteurised milk made nearly as many people sick as raw milk in the US, but that's because only a fraction of people were using raw milk. And it's hard to calculate the health benefits, because nobody has really worked on those numbers.

There is also an in-between step, called thermisation. You heat milk to 65°C (149°F), but only for 15 seconds. This kills 99.9% of dangerous *E. coli*, 99.9% of salmonella, and is 90% effective with staph. It is only 80% effective with listeria, but that's only an issue if you keep milk for more than three days.[18] This quick heating is better nutritively, and has little effect on many bioactive chemicals compared to pasteurisation. Maybe we could do this for raw milk?

Or do we continue with milk as moonshine, which people seek out like they once did hard liquor?

It's not just raw milk, it's all milk processing that is considered too hard by our bureaucrats. Where I live, it's now way easier to set up a distillery than it is a cheese factory, thanks to the regulatory burden. There are more than 80 whisky distilleries in Tasmania alone – about 10 times more distilleries than there are cheeseries. It seems odd that our priorities have pushed out cheesemaking as an option for small producers, when we know alcohol is toxic to every organ in the body, and milk is so nutritious.

Meanwhile, the trade in raw milk is booming – albeit on the black market.

Is it up to consumers to decide what risk they are willing to take, for the benefits, perceived or real, that they believe raw milk can provide? Do we trust the customers, or, like Timmermeister, do we wonder what they do with the stuff once it leaves the farm gate?

Of course, we would trust the customers if we had a legal product. We'd put on warning labels for pregnant women, the elderly, infants, those with compromised immune systems. We'd give it a three-day use-by date like the raw milk I bought once in France. That's why I think we should make it legal. Make it mandatory that people understand what the product is that they go to so much trouble to source. Build a bridge between those who want raw milk and good producers, and explain the risk. As someone who milks a cow, let me tell you it's a very, very complex interplay of things that makes milk safe, and most of them you can't see.

Milk isn't just pasteurised before you see it. It's pushed and pulled, from the feed of the cows, to the way it is heated, to the way it is 'standardised' and commodified. What we do to milk isn't always good for the milk, for our bodies, or for flavour. And that processing is what we're looking at next.

Hog Slop

Max Kregor is in his early 70s when he visits our farm. He was born in what was once the farmhouse for this property. That house still stands next to our driveway, though it now belongs to someone else. Max lived there while he grew up and stayed there until he got married, because oftentimes in those days people didn't move out of home until they were wed. So, he got married and moved away. Into the house next door.

When Max came to visit our farm, he knew a lot about the place. He told us it was all once orchard, except the parts where it is too steep to plant apple trees. He told us about the apple trees, planted before he was born, some of which still stand today. Max told us about the pigs his family had, because half the fruit they grew wasn't saleable – statistics mirrored, if not worse, with apples grown today. Apples with a stalk that's too short, or with flesh too pink, or with a patch of russet, could still be turned into something humans could and would eat, once it was put through a pig. Max talked about all of this. And he talked about his mum.

Max's mum didn't just cook for the household. Mrs Kregor cooked for the pickers, who filled every shed in autumn. And to cater, she ran a market garden, fattened pigs, and also milked three dairy cows.

Mrs Kregor used the cows to mow the grass between the orchard blocks, rather than tractors that burn fossil fuels. Cows grazed the parts of the property that were too steep or south-facing – areas that could only be used to grow grass. Or moss. Every morning she'd hand-milk the dairy cows, skim the cream off the top, churn the cream – Max remembers working the churn, having to spin it at just the right speed – and then Mrs Kregor would sell raw butter around the neighbourhood. But when she was too busy (and something tells me she was always busy), she would still hand-milk the cows. She would still skim off the cream, but instead of making butter, she'd put the cream churn out the front of the house, ready. Because, of course, she wasn't the only woman in the countryside milking dairy cows. Everybody in southern Tasmania who had any land milked one for themselves – and if space permitted, they'd milk three or more and sell the cream. A truck came down from the factory in Hobart, picked up the churns, and made butter to sell in town.

Now, Mrs Kregor didn't let her kids run short. They all drank tall glasses of full-cream milk. She took the cream and put it on scones, on porridge. She made custard. But three dairy cows produce a lot of milk. And that creates a problem. Every day she'd end up with 15–20 litres (3–4 gallons) of skimmed milk after making the butter or selling the cream. Some 15–20 litres of the low-fat stuff.

Remember, this is 1950s Tasmania. Mrs Kregor has fed her kids and all the farm workers. She's turned the parts of the farm that can't grow strawberries, or lentils, or soybeans, but can grow grass, into a food product, something humans can eat – milk – by using a ruminant and nothing more than sunlight. She's then transformed the cream into butter. But she then has a lot of skimmed milk to deal with. And these are God-fearing, hard-working people with a lot of self-respect. There's no way they or their neighbours would let anything they viewed as unfit for human consumption – in this case, skimmed milk – past their lips. So, there was a waste product.

Or was there? It turns out that rejected apples and market garden waste isn't the perfect diet for pigs, as ideally they need more protein.

But skimmed milk has milk's protein, without the fat – and together with the vitamins, minerals and roughage from fruit and vegetables, skimmed milk makes a far better food to help fatten a few porkers.

What happened, historically, on our farm is mirrored by what happened more generally. When the dairy industry grew to greater scale in the early 1900s, thanks to leaps in mechanisation, refrigeration and transport, people wanted more butter. And more cream. Before then, cream was only for those who had a cow, or bossed someone around who milked a cow. Suddenly, we could all get some.

But dairy cows only produce so much cream. And the more modern breeds, Holstein and Friesian, may yield more milk, but it's lower in butterfat. Our Jersey cows gift us milk that is up to 8.2% fat. Holsteins usually give about 3–4% fat. More milk with less fat is an issue when people want ghee and panna cotta and cheesecake. So, the dairy industry's solution was to find a home for skim milk.

Prior to World War II, skim milk was considered 'hog slop' in America as well as here in Australia. In fact, in the 1920s milk processors used to dump skim milk in rivers, until the US government banned it due to the smell.[1] In 1930, only 17% of the waste from Wisconsin's dairy processing plants was treated (reclaiming some of the protein for use in buttons, electrical insulators and glue)[2], leaving over 19,000 kg (42,000 lb) of whey, skim milk and buttermilk, which the factories dumped into waterways.[3]

The idea was born to actually sell the unsellable. Enter skim milk and its powder. US-made skim milk powder entered World War II in mess kitchens before American soldiers even hit the ground. Once the war ended, something once again had to be done with skim milk, and the powder made from it.

The timing was perfect for change. Consumers were being bombarded with messages about dairy being bad for heart health. People separated from the land didn't like the way cream would clot on top of their bottle from the supermarket. Suddenly, you could upsell the rubbish bit, the skim milk part, and keep the fat of the land, the cream, which already had its market share. A campaign by dairy manufacturers to sell low-fat milk worked, as did the saturated fat scare campaign. People flocked to lower-fat milks – and even today, US regulations still forbid the provision of full-fat milk in high schools (though a bill to change that did pass the House of Representatives federally in late 2023).

The new hog slop, sorry, skim milk was a diet food, marketed as a way to get your man! Kendra Smith-Howard, in her 2013 book *Pure and Modern Milk: An Environmental History Since 1900*, talks about the idea of skim milk as a diet food par excellence:

> Gayelord Hauser's *Live Younger, Live Longer*, which topped the best-sellers' list in 1951, called skim milk a 'wonder food'. Diets printed in women's magazines and even in Hoard's Dairyman recommended the food. A Sealtest skim milk offer gave consumers the chance to get a new bathroom scale with a proof-of-purchase tag.[4]

While this worked in the short term, we all know the consequences. People were turned off dairy. The milk they bought was thin and mean. The flavour was washed out. It didn't assuage their appetite like the real thing. It didn't help them lose weight. And the messaging brought all dairy into disrepute.

Of course, that led to the rise not of plant milk, but soft drink.

So where are we at today? Well, surprisingly, after half a century of fat-phobia, all over the Western world full-fat milk is enjoying something of a comeback, making up over half the sales of liquid milk in Australia. And while about 80–90% of the milk in developing countries is produced by smallholders, on patches of land under about 2 hectares (5 acres), and it's all used locally, many developed nations produce a surplus. In Australia, only 30% of the milk produced is drunk fresh; the rest is processed into cheese, butter and more.

Some milk is dried into powdered form, usually after being skimmed. Some is mined for its lactoferrin or casein or whey proteins, so body-builders can buy them in tins. Some is put into ultra-processed food.

But fresh milk? What you see in the shops is pretty much in the same form it came from the cow. Isn't it?

Well, no. Not entirely.

Milk isn't always pure milk these days. It's manipulated. From the genetics of the animal, to the hormones and supplements fed to the animal, to what happens once it leaves the farm. It's not just pasteurisation. Milk is often stripped apart. And it's often done in the name of consistency. After all, a single bottle of fresh milk could contain the milk from 5000 cows, all in different stages of lactation, or from different breeds.

If my cow is giving me milk that is 8.2% fat and 5.6% protein today, that will change as the seasons change. That will also change over her lifespan, and within each lactation. So, to avoid these natural variations, milk is 'standardised'. This means it has to match a chart that I imagine sits on the dairy processor's wall. It says, at least where I live, that milk should be at least 3.2% fat, and 3.0% protein. Skim milk should be a maximum of 0.15% fat and a minimum of 3.0% protein.

The aim, of course, is to stop people being short-changed – you certainly don't want lower than 3.2% butterfat if you've paid for whole milk. The way this is done in a factory, where milk comes from multiple farms and multiple breeds, and possibly from multiple days, is to pull it apart. It's forced through plates or a film, a form of microfiltration that separates the protein and some minerals and vitamins from the watery part that has dissolved salts and the sugars in it. There are two products in the most basic filtration, the permeate and the retentate. Permeate contains the lactose and many vitamins and minerals; the retentate has the protein (and the fat in the case of whole milk). Manufacturers can add permeate back to the retentate, or add it to skim milk if they like. That way they can put vitamins and minerals back into their milk and adjust the protein and fat to suit their needs, so long as they stick to those basic proportions.

Now, this seems harmless enough. But, ask any winemaker about what happens when you pump their grape juice, a fluid that is far simpler and less fragile than milk, and they'll tell you every step, every filter, every pump makes a difference. They know and care because their product is worth a lot more than $2 a litre, or whatever your milk costs now.

So, when milk is vigorously pulled apart and put back together, it's not the same as it originally was. I like to compare it to a jigsaw puzzle. Yes, the picture in the puzzle might be the same picture that was originally cut into pieces. But you can see the joins. Milk put back together resembles real milk – but just like a jigsaw puzzle, it's not the same. Yes, we can put the whole orange in our mouths, but we can't take it back out again and leave things as they were. Science has the ability to change things, but not always to change them back. We know that.

So, the milk you get has often been pulled apart and reassembled. It has also possibly been homogenised. This is a process of distributing the fat molecules differently. Milk straight from a cow will eventually form cream at the top (sheep and goat milk is more naturally homogenised). But some people don't like that the cream clots a bit when it rises, and it does also shorten the shelf life of milk, so usually milk is forced through plates or pipes, which breaks the fat globules into smaller parts. This does a few things.

One thing it does, obviously, is to stop the fat clumping again and cream forming at the top of your milk. But it does other less obvious things. Homogenisation changes the fat structure so much that it is absorbed differently in the body – more in the start of the gut and less later.[5] We're not really sure how that affects the body, especially after pasteurisation alters some of the proteins so they stick to the fat differently as well, but affect our digestion it does.

Homogenisation also changes the structure of milk visually. Dispersed micro-globules of butterfat make the milk look whiter. Less creamy. All this has an effect that is actually quite important. Homogenised milk has a different texture on the palate, it has less flavour because it doesn't linger in the mouth as long, and it has fewer aromatic qualities (also meaning it has less complexity, less taste if you will). Homogenise milk and you not only alter the fat composition, but lower the pleasure that milk is able to provide.

Milk as clothes

One of milk's wonderful qualities is its malleability. The way it transforms. Milk has even been used to make a fabric, Lanital, which was developed by Italian scientists prior to World War II, although the amount of casein needed to make production viable was unrealistic. At one stage it was used to replace fur in hats and to upholster car seats. None of which took off, thanks to the new plastics that were about to be made possible, and far more cheaply, with fossil fuels.

In the modern era of food technology, mucking about with milk is less about car seats and more about ultra-processed food and protein powders. This means milk is altered, and in ways that probably aren't so great. And remember, it all came from an animal, and all farming has a consequence, so if we're cutting down rainforests to keep dairy cows, or losing soil carbon to keep dairy cows, or being cruel to the animals themselves, then we might as well produce the caseins, lactose, fats and lactoferrin in a lab.

Skim milk and whey powder are both used in all kinds of places to boost protein content. And there are whole companies that focus on what is known as 'clean label' products – where, because they have isolated a milk ingredient, it can be put back into a dairy product without having to say it's there. All the ingredient label has to say is 'milk'.

So, 'milk' can include Promilk B-Max, a proprietary product from a company called Ingredia that has used ultra-processing technology to strip milk into parts ready to put back into other products, including milk itself.[6] Ingredia has the capacity to process

400 million litres (88 million gallons) of milk a year into subproducts. Promilk B-Max is packed with casein, which means you can make mousses without gelatine and cream cheese without gums.

Milk doesn't have this 'clean label' space to itself, of course. The same trick is used with plant-based foods, too, for everything from microfractions of rosemary to paprika, to give ultra-processed foods the perception of being natural when they're not.

Fresh milk, it must be said, is not considered ultra-processed even if it has been put through some filtration. Is it better than those milk substitutes that are an amalgam of stuff put together in a lab or a factory? Well, yes, according to all the data we have, all the reputable dietary advice from bodies such as the United Nations. Less is better when it comes to what is done to food in a processing plant, from milk all the way through to infant formula (which, by the way, is considered ultra-processed). It certainly doesn't hurt that milk is far more nutrient dense to begin with than any of its competitors.

In good news, many places offer milks for sale where little has been done to them – single-herd milks from small producers, perhaps using lower temperatures to pasteurise and leaving the milk unhomogenised. This milk is closer to the original; closer to best.

What makes milk great is its versatility. That quality – along with everything else it contains – is exceedingly hard to replicate in a factory.

For the next chapter, finally we get to leave the lab and the world of methane emissions and land use and ultra-processing behind. Now, we get to talk about what milk is particularly good at: making really, really delicious things to eat.

Blessed They Are

In the town of Bathmen in Holland, pig farmer Erik Stegink has been busy with coffee cups, two friends and a sow, collecting milk to make cheese. Because mother pigs only let down their milk for about 15 seconds and only give about a tablespoon per teat at a time, you really need to be there when they lactate, every 45 minutes, day and night.

It takes about 40 hours to get 10 litres (2 gallons) of pig milk using Erik's technique.[1] With a yield and labour bill like that, it's not surprising that we don't milk pigs much around the world. Pig cheese is even less likely, as the milk won't coagulate properly, apparently. Erik mixed his pig's milk with 80% cow's milk to make his *kaas*.

Erik is welcome to his reputation as arguably the only person in the world commercially milking pigs. While milking a sow is hard enough, dealing with mother pigs is always fraught. I had three trips to hospital, several stitches, batches of intravenous antibiotics that completely wiped my immune system, and weeks off work thanks to getting between a mother pig and her newborns. A tusk to the ankle will do that.

Cheesemaking itself is no easy feat. Yes, simple fresh cheeses are within the grasp of most people who cook, but matured cheeses – those with moulds or natural rinds that are drained, perhaps pressed, then aged – are quite tricky to get right.

In this chapter I want to explore what dairy can become. Milk is, to me, an amazing ingredient, partly because of how it is made out of other things by the animals that eat those things. And partly because of what it contains. And partly because of what it can become.

Milk's fat, sugar and protein can be transformed, often in a home kitchen. Let's start with milk.

Milk

Whole milk – or as I like to call it, 'milk' – has at least 3.2% fat and 3.0% protein if it's from a cow. Ideally it is unhomogenised (meaning the cream is still on the top and you have to shake it), because that tastes better and has better texture. It's also ideally pasteurised at a lower temperature than 65°C (149°F), to keep more of its true flavour and identity. Milk from a single herd on an organic or low-input farm, where the cows are grass fed and the milk is bottled locally, would be my ideal.

Low-fat milk can be anything below the 3.2% fat mark, and is usually around 1–2% butterfat. It's okay, but less than satisfying if you're used to the full flavour of milk. It still has the protein in it. Sometimes they add skim powder back to it, to give it more flavour.

Skim milk has below 0.15% fat. It's okay, I guess, if you have no self-respect. Again, they sometimes add skim milk powder or other concentrates back to some versions (which have different names) to give some semblance of flavour.

Yoghurt

The natural stuff is made at home by adding a small amount of a previous batch of natural yoghurt to milk. The milk is ideally heated to 92°C (198°F) first (just below boiling) for 15 minutes, because this changes the protein structure to create a thicker yoghurt. It's then cooled to about body temperature (just below 40°C/105°F), before about 1 tablespoon of yoghurt is added per litre of milk. This milk then needs to be incubated for at least six hours to let the yoghurt bacteria breed up and set the yoghurt. We have done this at home by pouring the cultured milk into sterilised jars, sealing them, then storing in an insulated container with water that is about 36°C (99°F) halfway up the side of the jars. Or using a yoghurt-making insulated plastic container. Once set, the yoghurt is better if left to sit for a couple of days in the fridge prior to using. It will then have a firm texture that cuts with a spoon, but will ooze a bit of loose whey. Commercially, to help prevent that, they sometimes add skim milk powder to the milk used, which ups the protein and sugar content.

Greek yoghurt is a creamy version of natural yoghurt. You add fresh cream to the milk you are making yoghurt from. About 10–20% cream is a nice addition.

Drained yoghurt became all the rage a few years ago because it gets rid of the whey that comes from natural yoghurt. You can simply put your natural yoghurt in a cloth-lined sieve and let it sit for a few hours in the fridge to thicken a bit. If you do this for a couple of days, it can become so thick that the yoghurt can be shaped into labneh, which stores better if rolled into balls and kept under oil in the fridge.

Strained yoghurt's popularity was born because most people find low-fat yoghurt thin and mean on the palate. Strained skim milk yoghurt concentrates the sugars and protein, so it is more satisfying.

Cream

Pouring cream, or single cream – or just 'cream' where I live – is about 35% butterfat. It's the top of the milk when it sits for a day or so after the cow is milked. At our place we usually skim it off the top of the pail with a ladle, but we also have a separator. Commercially, milk is skimmed using a separator that works like a centrifuge. Or it can go through filtration.

Double cream is about 45% fat, so it's thicker, richer.

Thickened cream is usually single cream (35% butterfat) with some kind of vegetable gum added to make it slightly less runny. It keeps for longer, but it's a lesser product than fresh cream, I think.

Sour cream is single cream that is cultured – using a bacteria, or a handful of different bacteria usually – in the same way that yoghurt is soured and set. Crème fraîche, the delectable French version, literally means 'fresh cream', but prior to refrigeration, it always soured slightly, as cream goes off so quickly. Crème fraîche is usually made using different species of bacteria to sour cream for a more complex flavour and sometimes looser texture.

Light (low-fat) sour cream is like sour cream, only not as nice. More vegetable gums are added to try to give it some kind of texture. It's ideal if you're allergic to flavour.

Clotted cream is made using double or even richer cream that is heated gently to caramelise the milk sugars that are still in it, and give it a sweet, baked character. It gets a crust on top, which is part of the joy of the eating – a brown, buttery, lightly crunchy part. Good clotted cream is ever so slightly soured. We make ours by pouring the thickest cream we have into a wide, flattish pan and putting it in a low oven (100°C/210°F) for a few hours. (If you like, you can place a water bath underneath the pan to keep it from overheating.) We scoop the clotted cream off the top, and make ice cream with the loose milk and cream underneath.

Cream cheese is more like a thick yoghurt. It's milk and cream, with a culture added (something that digests a bit of that lactose, such as lactic acid bacteria), an acid and salt. Commercially, they usually add a gum (such as locust bean gum) to thicken it. There are some good variations on the internet if you want to try making it from scratch.

Mascarpone is not a cheese, despite what some call it. It's a thickened cream from northern Italy that is great in desserts. Mascarpone can be made by adding an acid – such as tartaric acid, citric acid, or natural acidifiers such as vinegar or lemon or lime juice – to heated cream, then draining it a bit like you do for labneh. Homemade mascarpone is far better than most bought ones, I've found.

Butter

Oh wow. Butter. The thing margarine aspires to be when it grows up.

Butter is simply churned cream. Whip cream for long enough and you'll end up with the butterfat clumping together and the buttermilk falling out.

The cream for butter can be cultured, which – like yoghurt and sour cream – means using bacteria to add flavour complexity. One of the desired flavours is diacetyl, which some lactobacillus bacteria can produce. It adds caramel notes and the like (and is a chemical also made in factories to add a fake butter taste to margarine and popcorn). The bacteria also, of course, add a slight acidic note to the part of butter that's not fat, the buttermilk. Most of that is washed out in commercial butter, however.

There's sweet butter, salted butter and cultured butter. Sweet butter is simply churned cream. Salted butter, the type favoured in the UK, the US and Australia, is sweet butter that is washed in brine or has salt added. And cultured butter is as above. Though good cultured butter takes over a day to culture the cream, the best is then ripened for weeks in the fridge before being churned. Of course, chemical processes have allowed large butter manufacturers to do it in a matter of hours. The result isn't as nice.

Butter in the US is about 80% butterfat. In Europe, cultured butter must be at least 82% butterfat, and is usually higher than that.

In Australia it's about the same as the US, so it's less rich. This makes a difference in cooking, particularly for things like pastries. The part that isn't butterfat is milk solids, some buttermilk and water.

Butter can also have different textures depending on how it's churned and at what temperature. European butters are generally churned for longer, giving them a smaller fat crystal structure. It's kind of like the difference between chocolate that is ground for a short time and chocolate that is ground (conched) for a long time. Longer is better. It creates a creamier texture.

Clarified butter is normal butter that is heated, and then any stuff floating on top is skimmed off. Underneath is the melted butter, but you want to avoid the buttermilk and water, which will have sunk to the bottom. Ghee is cultured butter that is heated and clarified, so it is brought to the boil to cook out the buttermilk until it is all caramelised on the bottom of the pan, and all the other solids rise to the top and are skimmed off.

Cheese

Cheese is, as my friend and cheesemaker Nick Haddow puts it, milk immortalised. It's made by culturing milk, adding rennet, which is a clumping agent for proteins, and perhaps using salt and moulds and age.

Cultured milk with rennet added sets like a loose jelly and forms a soft curd. The more you cut or warm or move this curd, the more whey comes out. This curd is put in cheese moulds that are essentially shaped sieves that let the whey out, then they're turned regularly for a day or three. Most cheeses are the result of this culturing, setting and cutting process. Many others are also put under weights, to force out more whey, or pressed in other ways, some with blue or white moulds added, then ripened and matured.

The whey that comes out still has those whey proteins I talked about previously, and they can be recovered somewhat by heating the whey and adding an acid. (Usually the culturing process of cheese-making hasn't provided enough acid on its own.) This clumps the remaining proteins together, and the result is ricotta – fluffy light, white clouds. Ricotta, meaning 're-cooked' in Italian, is therefore a by-product of cheesemaking, and not technically a cheese. The acid is usually in the form of tartaric acid, though it can be vinegar, lemon juice or citric acid, too.

Some broad cheese varieties

Fresh and soft cheese

When cheese is first set, it's a jelly-like substance. This can be ladled off at this point, barely drained (particularly with richer milks such as buffalo, goat and sheep) and served as is, either sweet or savoury. You can also salt the curd and let it drain for a few days, or perhaps a week.

Sometimes mould is added to soft, fresh cheeses – particularly white moulds, usually variations on candida yeasts. These are usually added to the cheese when it's made, but they only breed where they have access to air, so the mould grows on the outside. Think of a brie or camembert or similar. Some white moulds create different folds in the cheeses, in particular geotrichum in goat's cheese, which creates a convoluted rind and a different taste. All these moulds work in a similar way, adding flavour while they break down the proteins from the outside of the cheese inwards. A gooey-edged goat crottin, for instance, or a runny farmhouse brie.

A variant on soft cheese is mozzarella, which is made by heating and stretching curds, to form balls of layered curd that trap some of the milky whey inside. The best mozzarella is made from buffalo milk and is an art perfected in Battipaglia and Caserta near Naples in southern Italy. It's at its best eaten immediately after making, and loses quality by the day.

Burrata is a cheese made in a similar way, but butter or cream is trapped inside the cheese. (People typically use cream these days.)

Stracciatella is a modern way of restaurants making a mozzarella-style cheese and folding cultured cream into it to serve. It's easier than trying to wrap cream inside molten mozzarella.

Squeaky cheese

Okay, so it's not really a formal category. But haloumi is a pretty popular cheese in the Mediterranean, and there are a few similar styles – including a so-called 'bread cheese' from Finland that I've tried, known as *leipäjuusto*.

These are cheeses that are cultured, cut and drained as per other cheeses, but after they're pressed, the fresh curds are then cut and simmered in whey. Technically they're not dissimilar to mozzarella, but they are firmer before being cooked, and so lend themselves well to barbecuing or frying.

Cooked curd cheese

Some marvellous semi-firm to hard cheeses were born in northern Italy, Switzerland, Austria, the French Alps and in the Basque regions. Think Gruyère, Emmental, Comté, Ossau Iraty and Piave. There's also the famed Manchego from central Spain and C2 from Bruny Island in Tassie. They're nutty, melting cheeses made by gently heating the curds of cow, sheep or goat milk to exclude more whey. Heating to 40–50°C (105–120°F), but not boiling. The most famous cooked curd cheeses are actually matured until hard, namely Grana Padano and Parmigiano Reggiano, both from Italy.

Hard cheese

As the curds drain, the cheese dries and you end up with a firmer result. These cheeses can be aged for many months, and the insides form glutamates, which have a highly satisfying umami effect on the palate. Think of the flavour of aged cheddar or a French Tomme and you'll get the picture.

Blue cheese

Not everybody's favourite, thanks to some pretty feisty cheap versions out there. The good ones have a creamy texture and sparkly acidity from green-blue mould (often *Penicillium roqueforti*). The curd is usually inoculated with the mould when made. Despite the myth that the mould follows the lines of copper wires that are poked into the cheese, it is simply an aerobic mould that needs air to grow, so it breeds up in the cheese along the holes left after it is spiked (with stainless steel pins). Blue cheese can be sweet and mild, or strong and acidic, and made from any milk you can make cheese with.

Washed rind cheese

This one frightens the horses. Pungent, smelly, sticky, these cheeses (usually with an orange coating of rind) are usually just a few weeks old and have been washed in brine, or wine, or beer or similar, gifting them a flavourful bacterium, usually *Brevibacterium linens*, which has a pungent smell not dissimilar to that of a teenager's socks. Some people, in my experience often cheesemakers, discard the rinds as that's where the most sulphurous compounds can sit. Washed rind cheeses include Bruny Island's 1770, Taleggio, Munster and Pont L'Eveque.

Vegetarian cheese

Traditionally – apart from a small number made using a rennet from thistles or artichokes – cheese was made with animal rennet, which is a coagulant enzyme found in the stomachs of young animals. To get it, the animal was killed, hence the problem vegetarians once had with eating cheese made with calf rennet or similar.

These days, most new-world cheesemakers use vegetarian rennet, which comes from GMO yeasts, just like some of that new lab dairy we're seeing. Some traditional 'vegetarian' cheeses from Europe still use animal rennet. If you're concerned, check the ingredients on the pack – though if you're really concerned about the unnecessary death of calves, you won't be eating any dairy.

Vegan cheese
Sorry, but it isn't cheese.

Is there anything vegan that might taste as complex as cheese?
The most interesting concoction that can resemble the variety you
see in cheese but isn't from an animal is tofu. From chou doufu,
the stinky tofu of Beijing, to the soft curd-like fresh tofu eaten as a
breakfast food with fried bread sticks in Malaysia, that's where much
cheese-like technique is best expressed. Soybean curds are cultured,
dried, smoked, pressed and have moulds applied, in ways that mirror
some of cheese's variety.

Answering the Big Questions

Why do we mostly milk cows, goats and sheep?

This one is pretty obvious when you think about it. Getting milk from an animal can be fraught. They might be dangerous (Tassie devils). They might not give much milk for the effort (pigs). The milk may not be to our tastes (seals, pigeons). We settled on mostly cows, buffalo, sheep and goats because we've managed to breed them to be docile enough not to kill or maim us, because they give a decent amount of milk for the effort, and because the milk they provide seems to fit our palate. And they also suit our purposes, both nutritionally and gastronomically, giving us products such as cheese.

We aren't baby cows, so why would we drink the milk of other animals?

Milk is the only food humans regularly consume that some of us have genetically mutated in recent millennia to be able to digest. You can't say that for chickpeas, or soybeans, or tomatoes or cabbage. It's a genetic mutation that has been rapidly taken up in the populations where it occurs, so it has to have been of benefit. So, are we supposed to drink the milk of other mammals? Well, nature has altered our very genetic code to allow some of us to be able to digest milk sugars, so it's probably fine, possibly good, and potentially great. If you don't want to drink it, or can't, that's fine, too.

Doesn't milk cause extra mucus?

The science on this is pretty clear. No, it doesn't cause nasal mucus (blocked sinuses, etc) from an allergic reaction. According to allergy.org.au:

> Studies have shown that milk has no effect on lung capacity, and does not trigger symptoms in patients with asthma. When people report coughing after having cold milk, it is usually due to breathing in cool air as they drink. This symptom generally disappears if the milk is warmed.

That said, people still, anecdotally, feel there is an increase in mucus production. One way it can possibly happen is reported in a 2010 scientific paper in a journal called *Medical Hypotheses*.[1] Researchers suggest it is through the way the proteins are broken down in the gut and isn't an allergy as we know it. We don't really know how it can happen, but they suggest a small subgroup of people may have good reason to believe dairy does increase mucus, and hence they also find benefit by eliminating dairy. If mucus is a problem for you, try it and make up your own mind.

I'm allergic to lactose

Well, you could be, but it's highly unlikely. What you may be is lactose intolerant. Allergies are a bit different, and most people who have dairy allergies are allergic to its proteins, not the sugar.

Allergies to casein and other major milk proteins are not hugely common, and can be tested for. If you've tried lactose-free milk and still get digestive problems from milk, the proteins could well be the cause. About half of all milk allergies (not intolerances) come from a reaction to beta-lactoglobulin, but scientists have discovered that the protein is only an allergen when not loaded with iron. The problem is, because milk is very low in iron (one of the essential nutrients it's very poor at providing), little research has been done into what helps increase milk's iron content, and more specifically, what attaches beta-lactoglobulin to it. What we do know is that diet and how cows are housed do affect iron levels. Organic milk, and the milk from animals that graze predominantly on fresh grass, has more iron. We also know that iron – and the allergic reaction to milk – is affected by what we do to milk, and that raw milk is better tolerated by some people. In fact, homogenisation, pasteurisation, and the cow's access to pasture (so not kept in sheds) all have an influence on the allergenic properties of milk – heating, homogenising and grain feeding making the milk more allergenic. Baked milk, which is milk that has been heated to a high temperature for a length of time (as happens in cakes etc), can be less allergenic, because the prolonged heating actually denatures (breaks apart) the proteins that some are allergic to. About 70% of children who are allergic to cow's milk can tolerate baked milk, so it's a handy tool, but obviously such heating reduces many of the other beneficial components of milk, too. The best way to tell what you're allergic or intolerant to is to get a doctor to do a test.

Are cheese dreams real?

While only a small fraction of people are actually allergic to milk proteins, they can be a bit trickier than some other proteins to digest. The suggestion is that cheese dreams, the broken sleep – and hence more vivid dream memories – after a late-night cheese binge could be a result of your body doing a bit more work than usual to break down your food.

How much lactose can a lactose-intolerant person actually consume?

If lactose-intolerant people can have some lactose, how much is enough? Apparently, individuals can have up to 25 grams (1 oz) at a sitting.[2] This would mean drinking 400–500 ml (about 2 cups) of fresh milk. That's a lot of milk. Some more recent science suggests it's safer to just have half that at a sitting,[3] but that still means people can digest 200 ml of milk at a time.

With yoghurt and fresh cheese, about half the lactose has already been digested, so lactose-intolerant people should, in theory, be able to eat 400 grams (14 oz) of fresh curd or yoghurt at a meal. Matured cheeses are virtually lactose free.

Can I become lactose intolerant later in life?

If you have the lactase persistence gene, then you should be able to safely consume lactose into adulthood. But some people who have drunk milk with no noticeable symptoms as grown-ups may find if they stop drinking milk for a while, they suffer some intolerance when they go back to consuming fresh dairy. The theory is that they aren't lactose tolerant, but their microbiome – gut bacteria – were breaking down the lactose in the diet for a while, but those bacteria dropped in number when no dairy was consumed. What that means is that those beneficial bugs aren't there in any great quantity to help you digest milk when you go back to it later in life. In theory, you should be able to gradually add dairy back to your diet to build up the lactose-eating biome.

Why do I get dots of butterfat on my tea from unhomogenised milk?

Homogenisation changes the structure of the fats, meaning the fat globules have about six times the surface area, and so don't clump together and won't form cream on the top of the milk, or dots of butter on top of your tea. Raw milk from a cow rarely does that either. But if you pasteurise unhomogenised milk, the butterfat clumps together more, making it denser – and if you heat the milk a bit too long or at too high a temperature, this clumping changes the cream so it will separate out in your tea. The clumping of the fat when heated is what makes clotted cream.

Those dots of butterfat on your tea are usually the result of over-zealous pasteurisation. It's a bit weird, and can alter the tea-drinking experience in a minor way, but it's not dangerous.

What is A2 milk, and is it better for you?

Hmm, well, this is super complicated. To put it simply, there's two types of one of the major milk proteins – a casein – and they differ by the cow. There are, in simple terms, A1 casein and A2 casein, and cows can produce a mix of the two, or pretty much just one or the other. We can test the cows for the type of protein they mainly produce, as it's genetically determined.

A2 doesn't show the potential for mucus production (see above). A1 protein slows down the passage of milk through the gut, and causes more inflammation. We don't actually know much of what that means, but we know it means something.

Some people report better dairy digestion from A2 milk, and some suggest (those promoting A2, mostly) that the A1 casein protein is responsible for many people's poor digestion of milk. Goats, buffalo and sheep produce predominantly A2 milk, as do certain breeds of cow, such as Jersey and Guernsey.

In short, if A2 milk gives you less digestive problems, then it could well be worth seeking out.

What are phytoestrogens, and will soy milk turn me from male to female?

Oestrogen is one of the major hormones that define what it means to be female biologically. It's involved in reproductive ability and it has very similar molecules, isoflavones, that exist in other things. There's also oestrogen in the milk we humans consume and buy, but in very low concentrations, and those hormones don't make it past the gut wall after the first day or so of our lives.

However, soy is one thing that is famed for having a compound similar to oestrogen – a plant oestrogen known as phytoestrogen.

Now, while isoflavones – phytoestrogens – can mimic oestrogen, they have no impact on male hormonal balance, according to a 2021 meta-analysis (big review of previous studies) in *Reproductive Toxicology*.[4] So, no, drinking soy milk won't affect a male's sperm count or libido. You can be just as much of a man drinking soy, or eating tofu.

Phytoestrogens do have some influence on female oestrogen levels, however: increasing the overall amount of oestrogen produced by the body in menopausal women, and slightly reducing it in women of ovulating age. This actually has some benefits, including reducing the incidence of breast cancer in the population.

Meanwhile, soybeans do have a bunch of other quite useful nutritional components, and soy milk is the least hollow, nutritionally, of the plant milks.

People with health issues, particularly cancers such as breast and prostate cancer, should consult a doctor about whether soy and/or milk are appropriate in their diet.

Is non-milk the milk answer?

It depends on the question. Milk produces (in many cases) fewer greenhouse gas emissions than red meat, and it's a high-quality protein that's important in some people's diet, but it's also not essential for anyone past the age of five.

Plant milks are pretty much all nutritionally bereft (apart from soy milk), but if you get your nutrition elsewhere, they can make your coffee whiter. Most travel a long way, and are packaged in seven layers of virgin materials to give them a long shelf life, which makes the cartons exceptionally hard, if not impossible, to recycle. Lab milks are fanciful future foods, and like plant milks are ultra-processed, which means you should limit them in your diet.

While nobody can scale up lab milks cost-effectively yet, their emissions aren't dissimilar to dairy, and the waste streams of both plant and cell milks are yet to be shown to be sustainable. You can certainly get all your nutrients without dairy after the age of natural weaning (over two years of age), and even for prior to that there are supplements that seem to help infants thrive.

So, is dairy the answer, or non-dairy? It depends on you, where you live, your moral compass and a whole bunch of cultural, gastronomic and societal factors.

Dairy isn't necessarily good or bad, but not all dairy is created equal. It certainly has a whole bunch of things in it – those micro RNAs, proteins, fats and EVs that seem to be of benefit. What researching this book taught me, and perhaps helped elucidate for you, is that real dairy is inordinately complicated in what it provides. It has benefits that we are only on the cusp of realising, from the very first drink from our mothers, through the cream we put on scones, to the cultured labneh in our salads.

It's been a blessing for humanity, embellishing culture, gracing tables, raising the gastronomic stakes and nourishing three-quarters of the world's population. It's the Swiss army knife of foods that has transformed our diets, and our genetic code, all while bringing pleasure to the lives of billions. And for that we should be grateful.

Acknowledgements

Firstly, thanks to the Fenton family for taking me on a bushwalk as a youngster over four decades ago to discover a farmhouse and their milking cow and to realise that dairy isn't just dairy. Thanks also to the women from Billawarra Dairy in Western Australia, who ran a two-cow yoghurt-making operation, for teaching me how to entice milk from recalcitrant cows. To Alan Benson, who made me question everything about how our food was produced, simply by doing a blind milk tasting over two decades ago. Kirsten Bradley led me down all kinds of milky paths in our chats around the kitchen bench, not the least when she told me about adoptive lactation. And to Nick Haddow, who has had so many milky conversations with me over the years, as well as a lot of good cheese.

Dairy people are very welcoming. I've been to countless farms over the years. Gratitude in particular to Rowan Jordan for showing me his organic dairy in Victoria, Naomi Steenkamp for giving me a tour of her goat creamery in New Zealand, Joey Malone and Jamie and Sheri Blackett for hosting me at Arbigland in Scotland, and Tom Gregory for being willing to share numbers on his carbon storage for his farm in England.

The team at Murdoch Books are the veritable cream – from the eagle eye of the best editor a writer could hope for, Katri Hilden, to the military precision of editorial management thanks to Justin Wolfers. From the enormous task of wrangling references by Bree Blundell to the unbridled enthusiasm for the topic from publicist Sue Bobbermein. Sue, I'll try not to get locked down with Covid in a hotel when touring for this book. And huge appreciation to publishing director Jane Morrow, who has backed me with multiple books that try to push the discussion on food; thanks for seeing the bigger picture. And for allowing those who actually grow and cook food to participate in the public debate of the how, why and what of food production globally. I reckon we've done some terrific things together and this book breaks totally new ground (or makes totally new curds?). Thanks for believing in me.

To Dr Daniella Susic, I'm indebted to you for bringing me up to speed on immunity and the vitality of human milk. A big shout out to Suzy Manigian, for spending so much time talking me through marsupial milks, and for being such a wildlife hero. Meanwhile, I gleaned so many insights into how hard it can be to feed our babies thanks to lactation consultant Louise Klug. I'm also supremely grateful to biologist Elise Ringwaldt for taking me out hunting devils on dark, cold, late-autumn Tasmanian mornings.

To Sadie, my love, thanks for allowing me the space to disappear into the office, or the muddy farm, yet again, to research and write, and for that all important first read through the manuscript. And thank you, too, for the meals, the laughs, the swims in alpine lakes that your enthusiasms bring to family life. I'll make some clotted cream and lemon curd ice cream in appreciation, because my love language is cooking.

A massive thumbs-up to the experts in this field: the scientists, especially those who did respond to my emails and requests. Just about all the research I used is open access, and you're welcome to peruse the reference list for just a taste of what is out there. Sadly, most of my requests for more detailed information from cell and plant milk companies went unanswered, a perhaps telling sign of how they would prefer to control the narrative.

And most of all, a thank you to you, the reader. I never imagined I'd ever write one book when I flunked Year 7 English, let alone 15 books, and I still have to pinch myself that others value my work enough to invest in it and read it. It's a privilege and an honour and a responsibility I don't take lightly, to contribute to our national and international conversations around the ethics, the source and the cultural appropriateness of what we put in our mouths.

Bibliography

Here is an abbreviated list of some reading around the topic that I found useful. A good introductory podcast is 'Gastropod, The Milk of Life', 21 June 2022.

Ehringer, G., *Leaving the Wild: The unnatural history of dogs, cats, cows and horses*, Pegasus Books, New York, 2017.

Freeman, A., *Skimmed: Breastfeeding, Race, and Injustice*, Stanford University Press, California, 2021.

Haddow, N., *Milk Made*, Hardie Grant, Melbourne, 2018.

Kessler, B., *Goat Song*, Simon & Schuster, New York, 2009.

Kurlansky, M., *Milk! A 10,000-year food fracas*, Bloomsbury Publishing, New York, 2018.

Meckel, R., *Save the Babies*, Boydell & Brewer, Suffolk, 1990 (and multiple reprints).

Smith-Howard, K., *Pure and Modern Milk: An environmental history since 1900*, Oxford University Press, Oxford, 2013.

Timmermeister, K., *Growing a Farmer: How I learned to live off the land*, W.W. Norton, New York, 2012.

Weaver, L.T., *White Blood: A history of human milk*, Unicorn Press, Lewes, East Sussex, 2021.

References

Chapter 1 – The Fat of the Land

1. Visioli, F. & Strata, A., 'Milk, dairy products, and their functional effects in humans: A narrative review of recent evidence', *Advances in Nutrition*, 1 March 2014, vol. 5 no. 2, pp. 131–43 <doi.org/10.3945/an.113.005025>

Chapter 2 – The Original Superfood

1. Weaver, L.T., 'Relationships between paediatricians and infant milk formula companies', *Archives of Disease in Childhood*, May 2006, vol. 91, no. 5, pp. 386–7 <doi.org/10.1136/adc.2004.069062>
2. Howarth, W.J., 'The influence of feeding on the mortality of infants', *The Lancet*, 22 July 1905, pp. 210–13 <ia800708.us.archive.org/view_archive.php?archive=/22/items/crossref-pre-1909-scholarly-works/10.1016%252Fs0140-6736%252800%252969299-1.zip&file=10.1016%252Fs0140-6736%252800%252969779-9.pdf>
3. Brady, J.P., 'Marketing breast milk substitutes: Problems and perils throughout the world', *Archives of Disease in Childhood*, 19 May 2012, vol. 97, no. 6, pp. 529–32 <doi.org/10.1136/archdischild-2011-301299>
4. Visioli, F. & Strata, A., 'Milk, dairy products, and their functional effects in humans: A narrative review of recent evidence', *Advances in Nutrition*, 1 March 2014, vol. 5, no. 2, pp. 131–43 <doi.org/10.3945/an.113.005025>
5. Whiting, A., 'Solving an ancient dairy mystery could help cure modern food ills', *The EU Research & Innovation Magazine*, European Commission, 21 January 2020 <ec.europa.eu/research-and-innovation/en/horizon-magazine/solving-ancient-dairy-mystery-could-help-cure-modern-food-ills>
6. '17 reasons to wean yourself from milk today', Animal Liberation <alv.org.au/cow-truth/17-reasons-to-wean-yourself-from-milk-today>
   &
   If you're looking at bacterial cells in nature, a single spinach leaf can have 600 different species of bacteria on it – at a rate of 10,000,000 microbes on a leaf. 'Tiny droplets allow bacteria to survive daytime dryness on leaves', *eLife*, 15 October 2019 <elifesciences.org/for-the-press/c0152f5b/tiny-droplets-allow-bacteria-to-survive-daytime-dryness-on-leaves>
7. Wickes, I.G., 'A history of infant feeding. Part III: Eighteenth and nineteenth century writers', *Archives of Disease in Childhood*, August 1953, vol. 28, no. 140, pp. 332–40 <doi.org/10.1136/adc.28.140.332>
8. Hardyment, C., 'Breast, bottle or goat's udder?', *The Guardian*, 18 November 2007 <theguardian.com/lifeandstyle/2007/nov/17/familyandrelationships.family2>

9. Männistö, S., Laatikainen, T., Helakorpi, S. et al., 'Monitoring diet and diet-related chronic disease risk factors in Finland', *Public Health Nutrition*, 1 June 2010, vol. 13, no. 6A, pp. 907–14 <doi.org/10.1017/S1368980010001084>

10. Severson, K., 'Got milk? Not this generation', *The New York Times*, 4 April 2023 <nytimes.com/2023/04/04/dining/milk-dairy-industry-gen-z.html>

11. Gumbel, A., 'Capitalism in a bottle', *Los Angeles Review of Books*, 19 November 2015 <lareviewofbooks.org/article/capitalism-in-a-bottle>

12. Stewart, H. & Kuchler, F., 'Fluid milk consumption continues downward trend, proving difficult to reverse', Economic Research Service, 21 June 2022 <ers.usda.gov/amber-waves/2022/june/fluid-milk-consumption-continues-downward-trend-proving-difficult-to-reverse>

13. Clarke, H., 'Change in UK consumer preferences show need for more cheese', Agriculture and Horticulture Development Board, 6 February 2020 <ahdb.org.uk/news/change-in-UK-consumer-preferences-for-more-cheese#:~:text=Average%20per%20capita%20consumption%20was,declined%20marginally%20over%20recent%20years>

Chapter 3 – The Devil's Drink

1. Landy, S., Peralta, S., Vogelnest, L. et al., 'The macroscopic and radiographic skull and dental pathology of the Tasmanian devil (*Sarcophilus harrisii*)', *Frontiers in Veterinary Science*, 10 June 2021, vol. 8 no. 693578, <doi.org/10.3389/fvets.2021.693578>

2. Landy et al., 'The macroscopic and radiographic skull … '

3. Di Silvestro, R., 'My, what a big bite you have', National Wildlife Federation, 1 December 2008 <nwf.org/Magazines/National-Wildlife/2009/Least-Weasel-Carnivore-Bites>

4. Schueman, L.J., 'Tasmanian devils: fierce predators with the strongest bite per body mass', *One Earth*, 28 April 2022 <oneearth.org/species-of-the-week-tasmanian-devil>

5. Karlen, S.J. & Krubitzer, L. (eds), 'Marsupial neocortex', in Squire, L.R. (ed.), *Encyclopedia of Neuroscience*, San Diego, CA: Academic Press, 2009, pp. 671–9 <doi.org/10.1016/B978-008045046-9.00966-9>
    &
    Stewart, F. & Tyndale-Biscoe, C.H. (eds), 'Pregnancy and parturition in marsupials', in Martini, L. & James, V.H.T. (eds), *The Endocrinology of Pregnancy and Parturition, Current Topics in Experimental Endocrinology*, vol. 4, New York, NY: Academic Press, 1983, pp. 1–33 <sciencedirect.com/science/article/abs/pii/B9780121532048500074>

6. 'Features of the honey possum', The University of Western Australia, 2011 <uwa.edu.au/study/-/media/Faculties/Science/Docs/Features-of-the-honey-possum.pdf>

Chapter 4 – Colostrum: Making Milk Look Like a Try-Hard

1. Harada, E. & Takeuchi, T., 'Possible role of colostral macromolecules transported from the intestinal lumen in neonatal animals', in Zabielski, R., Gregory, P.C., Weström, B. et al. (eds), *Biology of the Intestine in Growing Animals*, Amsterdam: Elsevier, 2002, pp. 465–89 <doi.org/10.1016/S1877-1823(09)70131-6>

2. Alamiri, F., Riesbeck, K. & Hakansson, A.P., 'HAMLET, a protein complex from human milk, has bactericidal activity and enhances the activity of antibiotics against pathogenic streptococci', *Antimicrobial Agents and Chemotherapy*, 21 November 2019, vol. 63, no. 12 <doi.org/10.1128/aac.01193-19>

3. Aydin, M.S., Yiğit, E.N., Vatandaşlar, E. et al., 'Transfer and integration of breast milk stem cells to the brain of suckling pups', *Scientific Reports*, 24 September 2018, vol. 8, no. 14289 <doi.org/10.1038/s41598-018-32715-5>

4. Kim, S.Y. & Yi, D.Y., 'Components of human breast milk: From macronutrient to microbiome and microRNA', *Clinical and Experimental Pediatrics*, 23 March 2020, vol. 63, no. 8, pp. 301–9 <doi.org/10.3345/cep.2020.00059>

Chapter 5 – Mutant Milk Drinkers

1. Gallo-Reynoso, J. & Ortiz, C.L., 'Feral cats steal milk from northern elephant seals', *Therya*, December 2010, vol. 1 no. 3, pp. 207–11 <scielo.org.mx/scielo.php?script=sci_arttext&pid=S2007-33642010000300005>

2. Petherick, A., 'Drink Milk, Be Merry?', International Milk Genomics Consortium, 2014 <milkgenomics.org/?splash=drink-milk-merry>

3. Cats can safely consume 6 g lactose/day (130 mls cow's milk, but allowing for variation, they recommend 85 mls/day). Beynen, A., 'Milk for cats', *Creature Companion*, May 2017, pp. 40–1 <doi.org/10.13140/RG.2.2.24040.06406>

4. Liebert, A., López, S., Jones, B.L. et al., 'World-wide distributions of lactase persistence alleles and the complex effects of recombination and selection', *Human Genetics*, 23 October 2017, vol. 136, pp. 1445–53 <doi.org/10.1007/s00439-017-1847-y>

5. Bintsis, T., 'Lactic acid bacteria as starter cultures: An update in their metabolism and genetics', *AIMS Microbiology*, 11 December 2018, vol. 4, no. 4, pp. 665–84 <doi.org/10.3934/microbiol.2018.4.665>

6. For the nerds, the gene variations are labelled as −13.910:T (combined with −22.018:A), in most of Eurasia, North Africa and Central Africa; −13.915:G, mostly in the Middle East; and −13.907:G, −14.009:G and −14.010:C, mostly in East Africa.

7. Bleasedale, M., Richter, K.K., Janzen, A. et al., 'Ancient proteins provide evidence of dairy consumption in eastern Africa', *Nature Communications*, 27 January 2021, vol. 12, no. 632 <doi.org/10.1038/s41467-020-20682-3>

8. Whiting, A., 'Solving an ancient dairy mystery could help cure modern food ills', *The EU Research & Innovation Magazine*, European Commission, 21 January 2020 <ec.europa.eu/research-and-innovation/en/horizon-magazine/solving-ancient-dairy-mystery-could-help-cure-modern-food-ills>

9. Whiting, 'Solving an ancient dairy mystery …'

10. Misselwitz, B., Butter, M., Verbeke, K. et al., 'Update on lactose malabsorption and intolerance: Pathogenesis, diagnosis and clinical management', *Gut*, 19 August 2019, vol. 68, no. 11, pp. 2080–91 <doi.org/10.1136/gutjnl-2019-318404>

11. Enserink, M., 'Did natural selection make the Dutch the tallest people on the planet?' *Science*, 7 April 2015, <science.org/content/article/did-natural-selection-make-dutch-tallest-people-planet>

Chapter 6 – Madonna's Breast

1. Weaver, L.T., *White Blood: A history of human milk*, Lewes, East Sussex: Unicorn Press, 2021, p. 42

2. Oftedal, O.T., 'Lactation in the dog: Milk composition and intake by puppies', *The Journal of Nutrition*, May 1984, vol. 114, no. 5, pp. 803–12 <doi.org/10.1093/jn/114.5.803>
&
Zhang, M., Sun, X., Cheng, J. et al., 'Analysis and comparison of nutrition profiles of canine milk with bovine and caprine milk', *Foods*, 5 February 2022, vol. 11 no. 3 <doi.org/10.3390/foods11030472>

3. Meyer, J.R., 'St. Bernard of Clairvaux', *Britannica*, 1 January 2024 <britannica.com/biography/Saint-Bernard-of-Clairvaux>

4. Downloaded from submissions to Australian Parliament House website. House of Representatives Committees, '(b) Evaluate the impact of marketing of breastmilk substitutes on breastfeeding rates and, in particular, in disadvantaged, Indigenous and remote communities', Parliament of Australia, pp. 31–2 <aph.gov.au/parliamentary_business/committees/house_of_representatives_committees?url=haa/breastfeeding/subs/sub269c.pdf>

5. Jackson, M. & Leslie, E., 'Deeper in the pyramid: Share of throat', Royal College of Art, 2023 <wellcomecollection.cdn.prismic.io/wellcomecollection/9811f41e-6d73-4ac0-9b12-4afc4778e846_ShareOfThroat-Accessible.pdf>

6. Stănescu, V., '"White power milk": Milk, dietary racism, and the "Alt-Right"', *Animal Studies Journal*, 2018, vol. 7 no. 2, pp. 103–28 <ro.uow.edu.au/cgi/viewcontent.cgi?article=1373&context=asj>

7. Cook, J., 'Got milk? How lactose tolerance influenced economic development', *PBS News Hour*, 3 December 2015 <pbs.org/newshour/economy/got-milk-lactose-tolerance-influenced-economic-development>

8. Baumgartel, K.L., Sneeringer, L. & Cohen, S.M., 'From royal wet nurses to Facebook: The evolution of breastmilk sharing', *Breastfeeding Review*, November 2016, vol. 24, no. 3, pp. 25–32 <ncbi.nlm.nih.gov/pmc/articles/PMC5603296>

9. During his time as Secretary of Commerce, Hoover did finally standardise milk bottle sizes, reducing the number of possibilities when buying milk from 32 different sizes to just four – so at least there's that.

Chapter 7 – Motherly Instinct

1. Newman, S., 'Infanticide', *Aeon*, 27 November 2017 <aeon.co/essays/the-roots-of-infanticide-run-deep-and-begin-with-poverty>
2. Garrod, B., 'How do whales breastfeed underwater?', Discover Wildlife, 4 May 2016 <discoverwildlife.com/animal-facts/marine-animals/how-do-whales-breastfeed-underwater>
3. Bittel, J., 'The strangest, most amazing lactation methods ever seen in mammals', *Slate*, 9 October 2015 <slate.com/technology/2015/10/lactation-in-mammals-humans-whales-seals-bats-and-echidnas.html>
4. Osterloff, E., 'The wonderfully weird world of tenrecs', Natural History Museum, 27 October 2022 <nhm.ac.uk/discover/the-weird-world-of-tenrecs.html#>
5. Galante, L., Milan, A.M., Reynolds, C.M. et al., 'Sex-specific human milk composition: The role of infant sex in determining early life nutrition', *Nutrients*, 1 September 2018, vol. 10, no. 9 <doi.org/10.3390/nu10091194>
6. Doucet, S., Soussignan, R., Sagot, P. et al., 'The secretion of areolar (Montgomery's) glands from lactating women elicits selective, unconditional responses in neonates', *PLoS One*, 23 October 2009, vol. 4, no. 10, article e7579 <doi.org/10.1371/journal.pone.0007579>
7. McBride, G., 'The "teat order" and communication in young pigs', *Animal Behaviour*, January 1963, vol. 11, no. 1, pp. 53–6 <doi.org/10.1016/0003-3472(63)90008-3>
   &
   Drake, A., Fraser, D. & Weary, D.M., 'Parent–offspring resource allocation in domestic pigs', *Behavioral Ecology and Sociobiology*, 2008, vol. 62, no. 3, pp. 309–19 <wellbeingintlstudiesrepository.org/cgi/viewcontent.cgi?article=1000&context=feebeh>
8. Malidaki, M. & Laska, M., 'Effects of an odor or taste stimulus applied to an artificial teat on the suckling behavior of newborn dairy calves', *Journal of Animal Science and Technology*, 16 April 2018, vol. 60, article 16, pp. 1–11 <doi.org/10.1186/s40781-018-0164-x>
9. 'Care of the newborn foal', Extension Foundation, 31 January 2020 <horses.extension.org/care-of-the-newborn-foal>
10. Beeton, I.M., *Mrs. Beeton's Book of Household Management*, London: Ward Locke, 1885 edition, p. 1069
11. World Health Organization, 'The physiological basis of breastfeeding', in World Health Organization, *Infant and Young Child Feeding: Model chapter for textbooks for medical students and allied health professionals*, Geneva: World Health Organization, 2009, p. 11 <ncbi.nlm.nih.gov/books/NBK148970>
12. Doucet et al., 'The secretion of areolar (Montgomery's) glands … '
13. The hormones around lactation also make the mother more aggressive. Studies show that lactating mothers can be more feisty, but with lower blood pressure; a cool, calm aggression that is a form of maternal protection, researchers believe.
14. Vasagar, J., 'Music hath charms to boost milk yield', *The Guardian*, 27 June 2001 <theguardian.com/uk/2001/jun/27/highereducation.research>
15. Ballard, O. & Morrow, A.L., 'Human milk composition: Nutrients and bioactive factors', *Pediatric Clinics of North America*, February 2013, vol. 60, no. 1, pp. 49–74 <doi.org/10.1016/j.pcl.2012.10.002>

Chapter 8 – The City Without Babies

1. Lannom, K.E. & Flowers, W.L., 'Teat location impacts colostrum and milk composition and piglet growth', National Hog Farmer, 24 July 2018 <nationalhogfarmer.com/animal-health/teat-location-impacts-colostrum-and-milk-composition-and-piglet-growth>
2. McCallum, N., 'US woman breastfed five puppies', *9News*, 11 February 2014 <9news.com.au/world/us-woman-breastfed-litter-of-five-puppies/663044bf-77ff-4ed0-9c77-07d6e706d7b3>
3. McLaren, J., *My Crowded Solitude*, London: Quality Press, 1926, p. 60
4. '1825: Toughen your nipples with puppies', Alpha History, 2023 <alphahistory.com/pastpeculiar/1825-toughen-your-nipples-puppies>
5. Shivram, B., 'Thicker than blood: The social and political significance of wet nurses in Mughal empire of North India', *Proceedings of the Indian History Congress*, 1 January 2008, vol. 69, pp. 403–16 <jstor.org/stable/44147204>
6. Akira, K., 'Kasuga no Tsubone and the establishment of the shōgun's inner chambers', Nippon, 22 July 2022 <nippon.com/en/japan-topics/c10801>
   &
   'Kitain temple', Kitain, 2021 <kitain.net/en>

7. Beeton, I.M., *Mrs. Beeton's Book of Household Management*, London: Ward Locke, 1885 edition, p. 1057

8. Beeton, *Mrs. Beeton's Book of Household Management*, pp. 1057–8

9. 'Breast is best? A history of wet nursing in France', Women in French History, 10 March 2022 <womeninfrenchhistory.com/an-interesting-history-of-wet-nursing-in-france>

10. 'Breast is best? … '

11. Dunn, P.M., 'Sir Hans Sloane (1660–1753) and the value of breast milk', *Archives of Disease in Childhood – Fatal and Neonatal Edition*, 1 July 2001, vol. 85, no. 1, pp. 73–4 <doi.org/10.1136/fn.85.1.F73>

12. Davenport, R., Schwarz, L. & Boulton, J., 'The decline of adult smallpox in eighteenth-century London', *The Economic History Review*, November 2011, vol. 64, no. 4, pp. 1289–1314 <doi.org/10.1111/j.1468-0289.2011.00599.x>

13. Kolovoz, B., 'Judge defends ejection of breastfeeding mother and baby from Melbourne court as "self-explanatory"', *The Guardian*, 10 March 2023 <theguardian.com/australia-news/2023/mar/10/judge-ejects-breastfeeding-mother-and-her-baby-from-melbourne-court-sparking-criticism>

14. Weaver, L.T., *White Blood: A history of human milk*, Lewes, East Sussex: Unicorn Press, 2021

15. Currier, R.W. & Widness, J.A., 'A brief history of milk hygiene and its impact on infant mortality from 1875 to 1925 and implications for today: A review', *Journal of Food Protection*, 1 October 2018, vol. 81, no. 10, pp. 1713–22 <doi.org/10.4315/0362-028X.JFP-18-186>

16. Stevens, E.E., Patrick, T.E. & Pickler, R., 'A history of infant feeding', *The Journal of Perinatal Education*, 1 January 2009, vol. 18, no. 2, pp. 32–9 <doi.org/10.1624/105812409X426314>

17. Calhoun, A.J., 'Slavery and racism drive a toxic double standard about breastfeeding', *The Washington Post*, 8 June 2022 <washingtonpost.com/outlook/2022/06/08/slavery-racism-drive-toxic-double-standard-about-breastfeeding>

18. Slatta, R.W., 'Newspaper advertisements for Black wet nurses (1821–1854), NC State University, 2019 <faculty.chass.ncsu.edu/slatta/hi216/documents/slavery/wetnurses.htm>

19. Slatta, 'Newspaper advertisements … '

20. Smith, J.P., Cohen, M. & Cassidy, T.M., 'Behind moves to regulate breastmilk trade lies the threat of a corporate takeover', *The Conversation*, 27 May 2021 <theconversation.com/behind-moves-to-regulate-breastmilk-trade-lies-the-threat-of-a-corporate-takeover-152446>

21. Farnsworth, S., 'Milk money: Indian company looking to sell breast milk to Australia', *ABC News*, 16 June 2017 <abc.net.au/news/2017-06-15/indian-company-neolacta-looking-to-sell-breast-milk-to-australia/8619020>

22. Steele, S., Foell, J., Martyn, J. et al., 'More than a lucrative liquid: The risks for adult consumers of human breast milk bought from the online market', *Journal of the Royal Society of Medicine*, 17 June 2015, vol. 108, no. 6, pp. 208–9 <doi.org/10.1177/0141076815588539>

Chapter 9 – Formulaic

1. ABC Australia, 'Tasmania, the roadkill capital of the world', YouTube, 29 October 2019 <youtube.com/watch?v=TsknzKfazac>

2. To put this in context, today's infant mortality rate is 2.72 in Australia, 3.94 in Canada, 5.48 in the US, 3.34 in the UK, and 3.1 in Europe more generally. Howarth, W.J., 'The influence of feeding on the mortality of infants', *The Lancet*, 22 July 1905, vol. 166, no. 4273, pp. 210–13 <ia800708.us.archive.org/view_archive.php?archive=/22/items/crossref-pre-1909-scholarly-works/10.1016%252Fs0140-6736%252800%252969299-1.zip&file=10.1016%252Fs0140-6736%252800%252969779-9.pdf>

3. Newmark, L.M., 'Residue of ruminant milk identified in prehistoric baby bottles', *Splash!*, 2019, no. 92 <milkgenomics.org/?splash=residue-of-ruminant-milk-identified-in-prehistoric-baby-bottles>

4. Shulman, S.T., 'The history of pediatric infectious diseases', *Pediatric Research*, 1 January 2004, vol. 55, pp. 163–76 <doi.org/10.1203/01.PDR.0000101756.93542.09>

5. Currier, R.W. & Widness, J.A., 'A brief history of milk hygiene and its impact on infant mortality from 1875 to 1925 and implications for today: A review', *Journal of Food Protection*, 1 October 2018, vol. 81, no. 10, pp. 1713–22 <doi.org/10.4315/0362-028X.JFP-18-186>

6. Fernandez, D., 'The surprisingly intolerant history of milk', *Smithsonian Magazine*, 11 May 2018 <smithsonianmag.com/history/surprisingly-intolerant-history-milk-180969056>

7. Meckel, R.A., *Save the Babies: American public health reform and the prevention of infant mortality, 1850–1929*, Rochester, NY: University of Rochester Press, 2015 edition <boydellandbrewer.com/9781580465175/save-the-babies>

8. 'Murder bottles: Baby feeding bottles that could kill', Baby Bottle Museum, 2016 <babybottle-museum. co.uk/murder-bottles>

9. Beeton, I.M., *Mrs. Beeton's Book of Household Management*, London: Ward Locke, 1885 edition

10. Atkins, P.J., 'Mother's milk and infant death in Britain, circa 1900-1940', *Anthropology of Food*, September 2003, vol. 2 <doi.org/10.4000/aof.310>

11. Morabia, A., Rubenstein, B. & Victora, C.G., 'Epidemiology and public health in 1906 England: Arthur Newsholme's methodological innovation to study breastfeeding and fatal diarrhea', *American Journal of Public Health*, July 2011, vol. 103, no. 7, pp. 17–22 <doi.org/10.2105/AJPH.2013.301227>

12. Atkins, 'Mother's milk and infant death … '

13. Stevens, E.E., Patrick, T.E. & Pickler, R., 'A history of infant feeding', *The Journal of Perinatal Education*, 1 January 2009, vol. 18, no. 2, pp. 32–9 <doi.org/10.1624/105812409X426314>

14. Lieffers, C., '"They perished in the cause of science": Justus von Liebig's food for infants', *Journal of the History of Medicine and Allied Sciences*, January 2024, vol. 79 no. 1, pp. 1–22 <doi.org/10.1093/jhmas/jrad035>

15. Tomori, C., 'New technologies claiming to copy human milk reuse old marketing tactics to sell baby formula and undermine breastfeeding', *The Conversation*, 14 June 2021 <news.yahoo.com/technologies-claiming-copy-human-milk-122451759.html?guccounter=1&guce_referrer=aHR0c HM6Ly93d3cuZ29vZ2xlLmNvbS88&guce_referrer_sig=AQAAALzy0GCCHzWqZeUutUK-MsLt5bqng 6ryxHPxJ2xZCanXHFgcNDhMlAbyGgIHQBYT1jIH4CUr1U7OyvVDf5LBJwDmQBOfXTuGBI6 QFGKge80vhHW4HviTPUdJgOVcnltf4-X91M3qn-jqRrVQg98O8i1yMAvo-N1QLgHlkbxiZKor>

16. Greenfield, B., 'The fascinating history of baby formula', Yahoo! Life, 24 August 2022 <au.news.yahoo. com/fascinating-history-of-baby-formula-231353714.html>

17. Holt, L.E., *The Care and Feeding of Children*, New York, NY: D. Appleton and Company, 1907 edition <gutenberg.org/files/15484/15484-h/15484-h.htm#Nursing>

18. Brady, J.P., 'Marketing breast milk substitutes: Problems and perils throughout the world', *Archives of Disease in Childhood*, 19 May 2012, vol. 97, no. 6, pp. 529–32 <doi.org/10.1136/ archdischild-2011-301299>

19. Nott, J., '"No one may starve in the British Empire": Kwashiorkor, protein and the politics of nutrition between Britain and Africa', *Social History of Medicine*, May 2021, vol. 34, no. 2, pp. 553–76 <doi. org/10.1093/shm/hkz107>

20. Currier, 'A brief history of milk hygiene … '

21. National Center for Chronic Disease Prevention and Health Promotion, 'Breastfeeding report card', Centers for Disease Control and Prevention, 2022 <cdc.gov/breastfeeding/data/reportcard.htm>

22. 'Breastfeeding rates in Australia', Australian Breastfeeding Association, May 2022 <breastfeeding.asn. au/resources/breastfeeding-rates-australia>

23. Formula is stipulated by FAO Codex. 'Standard for infant formula and formulas for special medical purposes intended for infants', The National Institute for Communicable Diseases, 2007 <nicd.ac.za/ wp-content/uploads/2018/05/Standard_for_Infant_Formula_and_Formulas_for_Special_Medical_ Purposes_intended_for_Infant_CODEX_STAN_72-1981_formerly_CAC_RS_72-1972.pdf> &
Breastmilk is here: Jenness, R., 'The composition of human milk', *Seminars in Perinatology*, July 1979, vol. 3 no. 3, pp. 225–39 <pubmed.ncbi.nlm.nih.gov/392766/#:~:text=Mature%20human%20milk%20 contains%203,colostrum%20than%20in%20mature%20milk> &
Colostrum for humans. Hester, S.N., Hustead, D.S., Mackey, A.D. et al., 'Is the macronutrient intake of formula-fed infants greater than breast-fed infants in early infancy?', *Journal of Nutrition and Metabolism*, 27 September 2012, vol. 2012, article 891201 <doi.org/10.1155/2012/891201> &
This works for cow's colostrum too. Polidori, P., Rapaccetti, R., Klimanova, Y. et al., 'Nutritional parameters in colostrum of different mammalian species', *Beverages*, 5 September 2022, vol. 8, no. 3, p. 54 <doi.org/10.3390/beverages8030054>

24. Schuman, A.J., 'A concise history of infant formula (twists and turns included)', *Contemporary Pediatrics*, 1 February 2003 <contemporarypediatrics.com/view/concise-history-infant-formula-twists-and-turns-included>

25. Mosley, T., 'How Black women were "skimmed" by infant formula marketing', WBUR, 16 December 2019 <wbur.org/hereandnow/2019/12/16/skimmed-black-women-formula-marketing>

26. Nott, '"No one may starve in the British Empire" … '

27. Muller, M., *The Baby Killer*, London: War on Want, 1974 <waronwant.org/sites/default/files/THE%20BABY%20KILLER%201974.pdf>

28. Department of Nutrition for Health and Development, 'The international code of marketing of breast-milk substitutes: Frequently asked questions', World Health Organization, 2017 <apps.who.int/iris/bitstream/handle/10665/254911/WHO-NMH-NHD-17.1-eng.pdf>

29. Cheung, K.Y., Petrou, L., Helfer, B. et al., 'Health and nutrition claims for infant formula: International cross sectional survey', *BMJ*, 15 February 2023, vol. 380, p. e071075 <doi.org/10.1136/bmj-2022-071075>

30. 'Addressing the infant formula shortage', The White House <whitehouse.gov/formula>

31. 'Public law 117–129: 117th Congress', Congressional Record, 21 May 2022, vol. 168 <govinfo.gov/content/pkg/PLAW-117publ129/pdf/PLAW-117publ129.pdf>

32. Yoaz, Y., 'Remedia execs, health officials face charges', World Alliance for Breastfeeding Action <waba.org.my/news/remedia_manslaughter.htm>

33. Fisher, G., 'Food technologist found guilty in 2003 baby formula deaths', *The Times of Israel*, 13 February 2013 <timesofisrael.com/food-technologist-found-guilty-in-2003-baby-formula-deaths>

34. Xiao, X., Loke, A.Y., Zhu, S. et al., '"The sweet and the bitter": mothers' experiences of breastfeeding in the early postpartum period: A qualitative exploratory study in China', *International Breastfeeding Journal*, 24 February 2020, vol. 15, article 12 <doi.org/10.1186/s13006-020-00256-1>

35. Tomori, 'New technologies claiming to copy human milk … '

36. Master, F., 'New rules set to shake up China's shrinking infant formula market', Reuters, 20 June 2023 <reuters.com/world/china/new-rules-set-shake-up-chinas-shrinking-infant-formula-market-2023-06-19>

37. Brady, 'Marketing breast milk substitutes … '

Chapter 10 – Don't Mind the Backwash

1. Vo, N., Tsai, T.C., Maxwell, C. et al., 'Early exposure to agricultural soil accelerates the maturation of the early-life pig gut microbiota', *Anaerobe*, June 2017, vol. 45, pp. 31–9 <doi.org/10.1016/j.anaerobe.2017.02.022>

2. Tooth, J., 'Why colostrum is so important for newborn pigs', Farmers Weekly, 12 July 2021 <fwi.co.uk/livestock/pigs/why-colostrum-is-so-important-for-newborn-pigs>

3. Inoue, R. & Tsukahara, T., 'Composition and physiological functions of the porcine colostrum', *Animal Science Journal*, 19 August 2021, vol. 92, no. 1, p. e13618 <doi.org/10.1111/asj.13618>

4. Durham, S.D., Wei, Z., Lemay, D.G. et al., 'Creation of a milk oligosaccharide database, MilkOligoDB, reveals common structural motifs and extensive diversity across mammals', *Scientific Reports*, 26 June 2023, vol. 13, article 10345 <doi.org/10.1038/s41598-023-36866-y>

5. Zhang, S., Chen, F., Zhang, Y. et al., 'Recent progress of porcine milk components and mammary gland function', *Journal of Animal Science and Biotechnology*, 22 October 2018, vol. 9, article 77 <doi.org/10.1186/s40104-018-0291-8>

6. Chen, W., Mi, J., Lv, N. et al., 'Lactation stage-dependency of the sow milk microbiota', *Frontiers in Microbiology*, 11 May 2018, vol. 9, no. 945 <doi.org/10.3389/fmicb.2018.00945>

7. Kim, S.Y. & Yi, D.Y., 'Components of human breast milk: From macronutrient to microbiome and microRNA', *Clinical and Experimental Pediatrics*, 23 March 2020, vol. 63, no. 8, pp. 301–9 <doi.org/10.3345/cep.2020.00059>

8. Khan, R., Petersen, F.C. & Shekhar, S., 'Commensal bacteria: An emerging player in defense against respiratory pathogens', *Frontiers in Immunology*, 31 May 2019, vol. 10, no. 1203 <doi.org/10.3389/fimmu.2019.01203>

9. Xu, R.J., Sangild, P.T., Zhang, Y.Q. et al., 'Chapter 5 Bioactive compounds in porcine colostrum and milk and their effects on intestinal development in neonatal pigs', *Biology of Growing Animals*, 2002, vol. 1, pp. 169–92 <doi.org/10.1016/S1877-1823(09)70121-3>

10. Rosa, F., Sharma, A.K., Gurung, M. et al., 'Human milk oligosaccharides impact cellular and inflammatory gene expression and immune response', *Frontiers in Immunology*, 29 June 2022, vol. 13, no. 907529 <doi.org/10.3389/fimmu.2022.907529>

11. Al-Shehri, S.S., Knox, C.L., Liley, H.G. et al., 'Breastmilk-saliva interactions boost innate immunity by regulating the oral microbiome in early infancy', *PLoS One*, 1 September 2015, vol. 10, no. 9, p. e0135047 <doi.org/10.1371/journal.pone.0135047>

12. Lyons, K.E., Ryan, C.A., Dempsey, E.M. et al., 'Breast milk, a source of beneficial microbes and associated benefits for infant health', *Nutrients*, 9 April 2020, vol. 12, no. 4 <doi.org/10.3390/nu12041039>

13. Hossain, S.T. & Barik, S., 'Why mothers kiss their babies? Does immune system play as a protective receptor of chemical determinants involved in mother-infant bonding?', *International Research Journal of Multidisciplinary Scope*, 2020, vol. 1, no. 2, pp. 38–40 <doi.org/10.47857/irjms.2020.v01i02.007>

14. Ghosh, S., Kumar, M., Santiana, M. et al., 'Enteric viruses replicate in salivary glands and infect through saliva', *Nature*, 29 June 2022, vol. 607, pp. 345–50 <doi.org/10.1038/s41586-022-04895-8>

15. Miklavcic, J.J., Badger, T.M., Bowlin, A.K. et al., 'Human breast-milk feeding enhances the humoral and cell-mediated immune response in neonatal piglets', *The Journal of Nutrition*, 1 November 2018, vol. 148, no. 11, pp. 1860–70 <doi.org/10.1093/jn/nxy170>

16. Nielsen, C.H., Hui, Y., Nguyen, D.N. et al., 'Alpha-lactalbumin enriched whey protein concentrate to improve gut, immunity and brain development in preterm pigs', *Nutrients*, 17 January 2020, vol. 12, no. 1, p. 245 <doi.org/10.3390/nu12010245>

17. 'Cytokines and their side effects', American Cancer Society, 27 December 2019 <cancer.org/cancer/managing-cancer/treatment-types/immunotherapy/cytokines.html#:~:text=Cytokines%20are%20small%20proteins%20that,body%27s%20immune%20and%20inflammation%20responses>

18. Niaz, B., Saeed, F., Ahmed, A. et al., 'Lactoferrin (LF): A natural antimicrobial protein', *International Journal of Food Properties*, 22 September 2019, vol. 22, no. 1, pp. 1626–41 <doi.org/10.1080/10942912.2019.1666137>

19. Boquien, C., 'Human milk: An ideal food for nutrition of preterm newborn', *Frontiers in Pediatrics*, 16 October 2018, vol. 6, no. 295 <doi.org/10.3389/fped.2018.00295>

20. Swelum, A.A., El-Saadony, M.T., Abdo, M. et al., 'Nutritional, antimicrobial and medicinal properties of camel's milk: A review', *Saudi Journal of Biological Sciences*, May 2021, vol. 28, no. 5, pp. 3126–36 <doi.org/10.1016/j.sjbs.2021.02.057>

21. 'Human lactoferrin "grown" in rice plants?', *Dairy Reporter*, 2 May 2002 <dairyreporter.com/Article/2002/05/02/Human-lactoferrin-grown-in-rice-plants>

Chapter 11 – How Real Milk is Made

1. Ashby, J., *Platypus Matters: The extraordinary story of Australian mammals*, Glasgow: William Collins, 2022

2. Bino, G., Kingsford, R.T., Archer, M. et al., 'The platypus: Evolutionary history, biology, and an uncertain future', *Journal of Mammalogy*, 24 April 2019, vol. 100, no. 2, pp. 308–27 <doi.org/10.1093/jmammal/gyz058>

3. 'Dolphin maternity', Dolphin Research Center <dolphins.org/maternity>

4. Arruda, L.B., Haider, N., Olayemi, A. et al., 'The niche of One Health approaches in Lassa fever surveillance and control', *Annals of Clinical Microbiology and Antimicrobials*, 24 April 2021, vol. 20, article 29 <doi.org/10.1186/s12941-021-00431-0>

5. Griffiths, M., McIntosh, D.L. & Leckie, R.M.C., 'The mammary glands of the red kangaroo with observations on the fatty acid components of the milk triglycerides', *Journal of Zoology*, February 1972, vol. 166, no. 2, pp. 265–75 <doi.org/10.1111/j.1469-7998.1972.tb04089.x>

6. Cortes, C., 'Blood supply' in Cortes, C., *Mammary Gland: Physiology and anatomy*, L'ecole Supérieure des Agriculture <groupe-esa.com/ladmec/bricks_modules/brick01/co/ZBO_Brick01_4.html>

7. Farmer, C., Trottier, N.L. & Dourmad, J.Y., 'Review: Current knowledge on mammary blood flow, mammary uptake of energetic precursors and their effects on sow milk yield', *Canadian Journal of Animal Science*, 1 June 2008, vol. 88, no. 2, pp. 195–204 <doi.org/10.4141/CJAS07074>

8. Geddes, D.T., Aljazaf, K.M., Kent, J.C. et al., 'Blood flow characteristics of the human lactating breast', *Journal of Human Lactation*, May 2012, vol. 28, no. 2, pp. 145–52 <pubmed.ncbi.nlm.nih.gov/22526342>

9. Geddes, D.T., 'Ultrasound imaging of the lactating breast: methodology and application', *International Breastfeeding Journal*, 29 April 2009, vol. 4, article 4 <doi.org/10.1186/1746-4358-4-4>

10. 'Amazing heart facts', Nova Online <pbs.org/wgbh/nova/heart/heartfacts.html>

11. Hassiotou, F. & Geddes, D.T., 'Immune cell–mediated protection of the mammary gland and the infant during breastfeeding', *Advances in Nutrition*, May 2015, vol. 6, no. 3, pp. 267–75 <doi.org/10.3945/an.114.007377>

12. Holm, M., Saraswat, M., Joenväärä, S. et al., 'Quantitative glycoproteomics of human milk and association with atopic disease', *PLoS One*, 13 May 2022, vol. 17, no. 5, p. e0267967 <doi.org/10.1371/journal.pone.0267967>

13. 'Mother's milk made to order for boys, girls', ABC Science, 17 February 2014 <abc.net.au/science/articles/2014/02/17/3946253.htm>

14. Paquette, A.F., Carbone, B.E., Vogel, S. et al., 'The human milk component *myo*-inositol promotes neuronal connectivity', *Proceedings of the National Academy of Sciences*, 11 July 2023, vol. 120, no. 30, p. e2221413120 <doi.org/10.1073/pnas.2221413120>

15. Witkowska-Zimny, M. & Kaminska-El-Hassan, E., 'Cells of human breast milk', *Cellular & Molecular Biology Letters*, 13 July 2017, vol. 22, article 11 <doi.org/10.1186/s11658-017-0042-4>

16. Nommsen, L.A., Lovelady, C.A., Heinig, M.J. et al., 'Determinants of energy, protein, lipid, and lactose concentrations in human milk during the first 12 mo of lactation: the DARLING Study', *The American Journal of Clinical Nutrition*, February 1991, vol. 53, no. 2, pp. 457–65 <doi.org/10.1093/ajcn/53.2.457>

17. Sollid, L.M., 'Breast milk against coeliac disease', *Gut*, 1 December 2002, vol. 51, no. 6, pp. 767–8 <doi.org/10.1136/gut.51.6.767>

18. Cohen, M., 'The lactating man', in Cohen, M. & Otomo, Y., *Making Milk: The past, present and future of our primary food*, London: Bloomsbury, 2017, pp. 141–60 <researchgate.net/publication/343079679_The_Lactating_Man>

19. Weimer, A.K., 'Lactation induction in a transgender woman: Macronutrient analysis and patient perspectives', *Journal of Human Lactation*, 3 May 2023, vol. 39, no. 3, pp. 488–94 <doi.org/10.1177/08903344231170559>

Chapter 12 – It's Not What You Know, It's What You Drink

1. Southey, F., 'Insect protein as beneficial as milk protein, find researchers', Food Navigator Europe, 8 June 2021 <foodnavigator.com/Article/2021/06/02/Insect-protein-as-beneficial-as-milk-protein-find-researchers>

2. White, O., 'Plant milk has been around for longer than you think', Vine Pair, 15 August 2022 <vinepair.com/articles/history-of-plant-milk>

3. Bridges, M., 'Moo-ove over, cow's milk: The rise of plant-based dairy alternatives', *Practical Gastroenterology*, January 2018, pp. 20–7 <med.virginia.edu/ginutrition/wp-content/uploads/sites/199/2014/06/January-18-Milk-Alternatives.pdf>

4. Mills, J., 'US milk and non-dairy milk market report 2023', Mintel <store.mintel.com/report/us-milk-and-non-dairy-milk-market-report>

5. Buttery, N., 'Mediaeval almond milk', *British Food: A History*, 29 April 2019 <britishfoodhistory.com/2019/04/29/mediaeval-almond-milk>

6. Suen, L., 'Erucic acid in edible fats and oils', *Food Safety Focus*, 16 January 2020, vol. 162 <cfs.gov.hk/english/multimedia/multimedia_pub/multimedia_pub_fsf_162_02.html>

7. Walther, B., Guggisberg, D., Badertscher, R. et al., 'Comparison of nutritional composition between plant-based drinks and cow's milk', *Frontiers in Nutrition*, 28 October 2022, vol. 9, p. e2645 <doi.org/10.3389/fnut.2022.988707>
&
Arrichiello, A., Auriemma, G. & Sarubbi, F., 'Comparison of nutritional value of different ruminant milks in human nutrition', *International Journal of Functional Nutrition*, July–August 2022, vol. 3, no. 4 <doi.org/10.3892/ijfn.2022.28>
&
This site has all the species listed, but doesn't do sugars, and their numbers are a bit different. Yasmin, I., Iqbal, R., Liaqat, A. et al., 'Characterization and comparative evaluation of milk protein variants from Pakistani dairy breeds', *Food Science of Animal Resources*, 1 September 2020, vol. 40, no. 5, pp. 689–98 <doi.org/10.5851/kosfa.2020.e44>

8. Swerling, G., 'Oat milk craze is fuelling vitamin deficiencies, experts fear', *The Telegraph*, 1 January 2024 <uk.news.yahoo.com/oat-milk-craze-fuelling-vitamin-163413156.html>
&
Godfrey, K.M., Titcombe, P., El-Heis, S. et al., 'Maternal B-vitamin and vitamin D status before, during, and after pregnancy and the influence of supplementation preconception and during pregnancy: Prespecified secondary analysis of the NiPPeR double-blind randomized controlled trial', *PLoS Medicine*, 5 December 2023, vol. 20, no. 12 <doi.org/10.1371/journal.pmed.1004260>

9. Muleya, M., Bailey, E.F. & Bailey, E.H., 'A comparison of the bioaccessible calcium supplies of various plant-based products relative to bovine milk', *Food Research International*, January 2024, vol. 175, p. e113795 <doi.org/10.1016/j.foodres.2023.113795>

10. Clay, N., Sexton, A.E., Garnett, T. et al., 'Palatable disruption: The politics of plant milk', *Agriculture and Human Values*, 30 January 2020, vol. 37, pp. 945–62 <doi.org/10.1007%2Fs10460-020-10022-y>
11. Stewart, H., 'Plant-based products replacing cow's milk, but the impact is small', Economic Research Service, 7 December 2020 <ers.usda.gov/amber-waves/2020/december/plant-based-products-replacing-cow-s-milk-but-the-impact-is-small>
12. Stewart, H. & Kuchler, F., 'Fluid milk consumption continues downward trend, proving difficult to reverse', Economic Research Service, 21 June 2022 <ers.usda.gov/amber-waves/2022/june/fluid-milk-consumption-continues-downward-trend-proving-difficult-to-reverse>
13. 'Liquid milk consumption per capita', Ibis World, 1 July 2023 <ibisworld.com/uk/bed/liquid-milk-consumption-per-capita/44115/>
14. Knight, B., 'Why plant-based "milks" are rising to the top', UNSW Sydney Newsroom, 28 July 2022 <newsroom.unsw.edu.au/news/health/why-plant-based-milks-are-rising-top>
15. 'How does milk compare to plant-based beverages?', Dairy Australia, May 2022 <cdn-prod.dairyaustralia.com.au/-/media/dairy/files/sustainability/sustainable-diets/dairy-and-plant-based-beverages-infographic-2022.pdf?rev=ee704c174b0044b7bbfb0a7ee1ae4547&hash=F84B385F3680967B6CC4208ACF0B4A54>
16. Clay et al., 'Palatable disruption … '

Chapter 13 – Cows With Guns

1. Eliasson, S., 'The 10 biggest farms in the world', *AgronoMag*, 19 April 2023 <agronomag.com/top-10-biggest-farms-world>
2. CGTN, 'Feeding 1.4 billion: Inside China's largest dairy farm', YouTube, 9 November 2019 <youtube.com/watch?v=QnNVX7ViOZU>
3. Southern, K., 'Texas dairy farm explosion kills 18,000 cows', *The Times*, 14 April 2023 <thetimes.co.uk/article/texas-dairy-farm-explosion-kills-thousands-of-cows-7mtg8790j>
4. Gentle, T., 'Moxey Farms 360° video tour – Australia's largest single site dairy farming operation', YouTube, 6 December 2016 <youtube.com/watch?v=5RWSwSJ6smM>
5. 'Switchboard for Moxey Farms tunnel barns 8 and 9', Total Electrical Control Solutions, March 2019 <tec-solutions.com.au/project/switchboard-moxey-farms-tunnel-barns-8-and-9>
6. 'Operations', Australian Fresh Milk Holdings, 2024 <afmh.com.au/operations>
7. Donohoe, P., 'In pictures: Inside Almarai's dairy farm in the Saudi desert', *Irish Farmers Journal*, 28 February 2017 <farmersjournal.ie/in-pictures-inside-almarais-dairy-farm-in-the-saudi-desert-258662>
8. Wakeham, K., Jarenwattananon, P. & Summers, J., 'Amid a water crisis, Arizona is using lots of it to grow alfalfa to export overseas', *WESA*, 9 August 2023 <wesa.fm/national-international-news/national-international-news/2023-08-09/amid-a-water-crisis-arizona-is-using-lots-of-it-to-grow-alfalfa-to-export-overseas>
   &
   Mechanic, M., 'US groundwater is being shipped overseas', *Mother Jones*, 13 December 2023 <motherjones.com/politics/2023/12/groundwater-us-exports-alfalfa-hay-china-saudi-arabia-united-arab-emirates-arizona>
9. Garvey, M., 'Lameness in dairy cow herds: Disease aetiology, prevention and management', *Dairy*, 18 March 2022, vol. 3, no. 1, pp. 199–210 <doi.org/10.3390/dairy3010016>
10. Fleischer, P., Metzner, M., Beyerbach, M. et al., 'The relationship between milk yield and the incidence of some diseases in dairy cows', *Journal of Dairy Science*, 2001, vol. 84, no. 9, pp. 2025–35 <doi.org/10.3168/jds.S0022-0302(01)74646-2>
11. Peterson, C.B. & Mitloehner, F.M., 'Sustainability of the dairy industry: Emissions and mitigation opportunities', *Frontiers in Animal Science*, 18 October 2021, vol. 2, article 760310 <doi.org/10.3389/fanim.2021.760310>
12. Mac, S.E., Lomax, S. & Clark, C.E.F., 'Behavioral responses to cow and calf separation: Separation at 1 and 100 days after birth', *Animal Bioscience*, May 2023, vol. 36, no. 5, pp. 810–17 <doi.org/10.5713/ab.22.0257>
13. Weary, D.M. & Chua, B., 'Effects of early separation on the dairy cow and calf: 1. Separation at 6 h, 1 day and 4 days after birth', *Applied Animal Behaviour Science*, 1 October 2000, vol. 69, no. 3, pp. 177–88 <doi.org/10.1016/S0168-1591(00)00128-3>

14. Griffiths, J., *What's For Dinner? Our food, our choices, our planet*, Melbourne: Thames & Hudson, 2023

15. Nicolao, A., Veissier, I., Bouchon, M. et al., 'Animal performance and stress at weaning when dairy cows suckle their calves for short versus long daily durations', *Animal*, June 2022, vol. 16, no. 6, p. e100536 <doi.org/10.1016/j.animal.2022.100536>

16. Dohoo, I.R., DesCôteaux, L., Leslie, K. et al., 'A meta-analysis review of the effects of recombinant bovine somatotropin', *The Canadian Journal of Veterinary Research*, October 2003, vol. 67, no. 4, pp. 252–64 <ncbi.nlm.nih.gov/pmc/articles/PMC280709>

17. Erasmus, L.-M., van Marle-Köster, E., Masenge, A. et al., 'Exploring the effect of auditory stimuli on activity levels, milk yield and faecal glucocorticoid metabolite concentrations in Holstein cows', *Domestic Animal Endocrinology*, January 2023, vol. 82, p. e106767 <doi.org/10.1016/j.domaniend.2022.106767>
    &
    University of Leicester, 'Milk yields affected by music tempo', *Biology Online*, June 2001 <biologyonline.com/articles/milk-yields-affected-music>

18. Joy, M., personal correspondence, August 2023

19. Melhelm, Y.B., 'New Zealand's troubled waters', *ABC News*, 17 March 2021 <abc.net.au/news/2021-03-16/new-zealand-rivers-pollution-100-per-cent-pure/13236174>

20. Silva, B.Q. & Smetana, S., 'Review on milk substitutes from an environmental and nutritional point of view', *Applied Food Research*, June 2022, vol. 2, no. 1, p. e100105 <doi.org/10.1016/j.afres.2022.100105>

21. 'Arable land (% of land area) –Australia', The World Bank <data.worldbank.org/indicator/AG.LND.ARBL.ZS?locations=AU>

22. '7. Dairy and dairy products', in *OECD-FAO Agricultural Outlook 2020-2029*, Paris: OECD Publishing, 2020 <oecd-ilibrary.org/sites/aa3fa6a0-en/index.html?itemId=/content/component/aa3fa6a0-en>

23. 'Methane emissions from the energy sector are 70% higher than official figures', International Energy Agency, 23 February 2022 <iea.org/news/methane-emissions-from-the-energy-sector-are-70-higher-than-official-figures>

24. Sherwin, E.D., Rutherford, J.S., Zhang, Z. et al., 'US oil and gas system emissions from nearly one million aerial site measurements', *Nature*, 13 March 2024, vol. 627 no. 8003, pp. 328–34 <doi.org/10.1038/s41586-024-07117-5>

25. Ivanovich, C.C., Sun, T., Gordon, D.R. et al., 'Future warming from global food consumption', *Nature Climate Change*, 6 March 2023, vol. 13, pp. 297–302 <doi.org/10.1038/s41558-023-01605-8>

26. Grant, C.A. & Hicks, A.L., 'Comparative life cycle assessment of milk and plant-based alternatives', *Environmental Engineering Science*, 5 November 2018, vol. 35, no. 11, pp. 1235–47 <doi.org/10.1089/ees.2018.0233>
    &
    Silva et al., 'Review on milk … '
    &
    'Total methane production in milk production reduced by 57 per cent in 60 years – further reductions still possible', Valio, 15 June 2022 <valio.com/news/total-methane-production-in-milk-production-reduced-by-57-per-cent-in-60-years-further-reductions-still-possible>

27. US, methane is 12% of all human GHG emissions. Enteric fermentation is 25% of that, so just 3% of total. Landfill is 15% of methane, 35% is coal and gas emissions. 'Greenhouse gas emissions: Overview of greenhouse gases', United States Environmental Protection Agency, 10 October 2023 <epa.gov/ghgemissions/overview-greenhouse-gases#methane>
    &
    Dairy is 23% of total enteric. So, dairy enteric methane is about 0.7% of GHG emissions.
    Segers, J.R. & Knox, P., 'What's the "beef" with methane emission and cattle production?', UGA Cooperative Extension Bulletin 1453, 1 December 2022 <extension.uga.edu/publications/detail.html?number=B1453&title=whats-the-beef-with-methane-emission-and-cattle-production>

28. 'What is the Australian dairy industry's emissions reduction target?', Dairy Australia, 17 January 2024 <dairy.com.au/dairy-matters/you-ask-we-answer/yawa-187---what-is-the-australian-dairy-industrys-emissions-reduction-target>

29. China Resources website, <en.crc.com.cn/whatwedo/power/>

30. Dwyer, O., 'Food waste makes up "half" of global food system emissions', Carbon Brief, 13 March 2023 <carbonbrief.org/food-waste-makes-up-half-of-global-food-system-emissions>

31. One paper coffee cup emits at least 110 grams (4 oz) of CO_2 (and more with the lid). Morgan, A., 'Paper cup vs mug – what's better for the environment?', EcoRate, 11 May 2022 <ecorate.eco/blog/paper-cup-vs-mug-whats-better-for-the-environment>
&
Raghavan, R. & Notaras, M., 'Sad demise of the paper coffee cup', Our World, 3 March 2009 <ourworld.unu.edu/en/storm-in-a-paper-cup>
&
Nowell, C., 'The disposable cup crisis: What's the environmental impact of a to-go coffee?', *The Guardian*, 23 January 2024 <theguardian.com/environment/2024/jan/22/disposable-coffee-cups-environmental-impact>
&
Roughly 1 kg (2 lb 4 oz) CO_2 equivalent for a litre of milk on a pasture-based farm. 'Greenhouse gas accounting: Australian dairy carbon calculator', Dairy Australia, 2016 <cdn-prod.dairyaustralia.com.au/-/media/project/dairy-australia-sites/national-home/resources/2020/07/08/dairy-carbon-calculator-factsheet/dairy-carbon-calculator-factsheet.pdf>
&
About 120 ml (4 fl oz) milk in a caffe latte. 'How many calories are in coffee?', Anytime Fitness Australia, 16 May 2019 <anytimefitness.com.au/blog/nutrition-recipes/how-many-calories-in-coffee>
&
And a flat white. Sawyer, G., 'Aussie flat white coffee 101: Your definitive guide', Coffeefusion, 10 August 2022 <www.coffeefusion.com.au/blogs/news/flat-white-coffee>
32. Rainer, E.M., Seppey, C.V.W., Tveit, A.T. et al., 'Methanotroph populations and CH_4 oxidation potentials in high-Arctic peat are altered by herbivory induced vegetation change', *FEMS Microbiology Ecology*, October 2020, vol. 96, no. 10 <doi.org/10.1093/femsec/fiaa140>
33. Nickel, R., 'The climate-friendly cows bred to belch less methane', Reuters, 9 August 2023 <reuters.com/business/environment/climate-friendly-cows-bred-belch-less-methane-2023-08-08>
34. Cheng, L., Zhang, X., Reis, S. et al., 'A 12% switch from monogastric to ruminant livestock production can reduce emissions and boost crop production for 525 million people', *Nature Food*, vol. 3, pp. 1040–51 <doi.org/10.1038/s43016-022-00661-1>
35. Miller, E., 'Rebel Kitchen brings Barista Mylk to Australia', *BeanScene Magazine*, 3 May 2021 <beanscenemag.com.au/rebel-kitchen-brings-barista-mylk-to-australia>
36. Naughton, J., 'Why AI is a disaster for the climate', *The Guardian*, 24 December 2023 <theguardian.com/commentisfree/2023/dec/23/ai-chat-gpt-environmental-impact-energy-carbon-intensive-technology>
37. Nanavatty, R., 'Air conditioning is threatening our ability to tackle climate change. Here's what we need to do', World Economic Forum, 10 January 2019 <weforum.org/agenda/2019/01/why-keeping-ourselves-cool-doesnt-have-to-mean-heating-the-planet>
38. Temkin, A.M., Evans, S., Spyropoulos, D.D. et al., 'A pilot study of chlormequat in food and urine from adults in the United States from 2017 to 2023', *Journal of Exposure Science & Environmental Epidemiology*, 15 February 2024 <doi.org/10.1038/s41370-024-00643-4>

Chapter 14 – Don't Have a Cow!

1. Ding, J., Liao, N., Zheng, Y. et al., 'The composition and function of pigeon milk microbiota transmitted from parent pigeons to squabs', *Frontiers in Microbiology*, 4 August 2020, vol. 11 <doi.org/10.3389/fmicb.2020.01789>
2. Fan, H. & Duncan, M., 'Cows churn out "human breastmilk"', Reuters, 17 June 2011 <reuters.com/article/us-china-cows-idUSTRE75F10K20110616>
3. Thomson, A., 'Lab-grown milk to hit shelves by 2024 – minus the cow and the carbon', *The Age*, 13 September 2022 <theage.com.au/business/entrepreneurship/lab-grown-milk-to-hit-shelves-by-2024-minus-the-cow-and-the-carbon-20220905-p5bfji.html?btis>
4. Lu, D., 'Leading the whey: The synthetic milk startups shaking up the dairy industry', *The Guardian*, 18 September 2022 <theguardian.com/food/2022/sep/18/leading-the-whey-the-synthetic-milk-startups-shaking-up-the-dairy-industry>
5. Very Dairy Singapore <facebook.com/verydairysg>
6. 'General Mills launches Bold Cultr Cream Cheese made with remilk animal-free dairy', *Vegconomist*, 16 January 2023 <vegconomist.com/products-launches/general-mills-bold-cultr-remilk>
7. The Kitchen FoodTech Hub <thekitchenhub.com>

8. Bhutada, G., Menard, G., Bhunia, R.K. et al., 'Production of human milk fat substitute by engineered strains of *Yarrowia lipolytica*', *Metabolic Engineering Communications*, June 2022, vol. 14, p. e00192 <doi.org/10.1016/j.mec.2022.e00192>

9. Hettinga, K. & Bijl, E., 'Can recombinant milk proteins replace those produced by animals?', *Current Opinion in Biotechnology*, June 2022, vol. 75, no. 102690 <doi.org/10.1016/j.copbio.2022.102690>

10. Delosière, M., Pires, J.A.A., Bernard, L. et al., 'Dataset reporting 4654 cow milk proteins listed according to lactation stages and milk fractions', *Data in Brief*, April 2020, vol. 29, no. 105105 <doi.org/10.1016/j.dib.2019.105105>

11. Behm, K., Nappa, M., Aro, N. et al., 'Comparison of carbon footprint and water scarcity footprint of milk protein produced by cellular agriculture and the dairy industry', *The International Journal of Life Cycle Assessment*, 26 August 2022, vol. 27, pp. 1017–34 <doi.org/10.1007/s11367-022-02087-0>

Chapter 15 – Love Oatmilk? Good. Someone Better Love Bacon

1. Banerjee, S., Coussens, N.P., Gallat, F. et al., 'Structure of a heterogeneous, glycosylated, lipid-bound, *in vivo*-grown protein crystal at atomic resolution from the viviparous cockroach *Diploptera punctata*', *IUCrJ*, July 2016, vol. 3, no. 4, pp. 282–93 <doi.org/10.1107/S2052252516008903>

2. Bowler, J., 'Scientists think cockroach milk could be the next superfood, and we wish we were kidding', *Science Alert*, 7 April 2018 <sciencealert.com/scientists-think-we-should-start-drinking-cockroach-milk-superfood>

3. Sidenmark, J., 'Over & oat – Turning oat drink residues into new food and materials', Axfoundation, 2024 <axfoundation.se/en/projects/over-n-oat>

4. McKee, D., 'Oats in high demand for milk production', World-Grain.com, 13 February 2023 <world-grain.com/articles/18099-oats-in-high-demand-for-milk-production>

5. 'Oatly and our fiber residues', Oatly, 8 September 2022 <community.oatly.com/conversations/news-and-views/oatly-and-our-fiber-residues/6318b759eb08200ed8a11f96>

6. 'The Oatly Sustainability Update 2022', page 27 <a.storyblok.com/f/107921/x/11daa2b42e/oatly-sustainability-report-2022.pdf#page=28>

7. O'Toole, D.K., 'Characteristics and use of okara, the soybean residue from soy milk production – a review', *Journal of Agricultural and Food Chemistry*, 29 January 1999, vol. 47, no. 2, pp. 363–71 <pubs.acs.org/doi/10.1021/jf980754l>
&
Li, B., Qiao, M. & Lu, F., 'Composition, nutrition, and utilization of okara (soybean residue)', *Food Reviews International*, 24 April 2012, vol. 28, no. 3, pp. 231–52 <doi.org/10.1080/87559129.2011.595023>

8. 'Everything you need to know about okara (but were afraid to ask)', Tetra Pak <tetrapak.com/insights/cases-articles/everything-you-need-to-know-about-okara>

9. Kamble, D.B. & Rani, S., 'Bioactive components, in vitro digestibility, microstructure and application of soybean residue (okara): A review', *Legume Science*, March 2020, vol. 2, no. 1, p. e32 <doi.org/10.1002/leg3.32>

10. Barral-Martinez, M., Fraga-Corral, M., Garcia-Perez, P. et al., 'Almond by-products: Valorization for sustainability and competitiveness of the industry', *Foods*, 2 August 2021, vol. 10, no. 8, p. e1793 <doi.org/10.3390/foods10081793>

11. Jones, M.H., 'Finding a greater use of almond byproducts', *AG Daily*, 25 July 2019 <agdaily.com/crops/almonds-alternative-uses>

12. Pardo-Giménez, A., Carrasco, J., Roncero, J.M.M et al., 'Recycling of the biomass waste defatted almond meal as a novel nutritional supplementation for cultivated edible mushrooms', *Acta Scientiarum Agronomy*, 15 May 2018, vol. 40, no. 1, p. e39341 <doi.org/10.4025/actasciagron.v40i1.39341>

13. Staight, K., 'Biochar industry fuelled by agricultural waste expected to grow', *ABC Rural*, 1 October 2022 <abc.net.au/news/rural/2022-10-01/biochar-industry-grows-in-australia-big-benefits-for-agriculture/101483868>

14. Bountiful, 'Almond water footprint, a new perspective', *Medium*, 23 April 2020 <medium.com/@BountifulAg/almond-water-footprint-a-new-perspective-7e401851d344>

15. Naylor, T., 'Ditch the almond milk: Why everything you know about sustainable eating is probably wrong', *The Guardian*, 5 September 2018 <theguardian.com/food/2018/sep/05/ditch-the-almond-milk-why-everything-you-know-about-sustainable-eating-is-probably-wrong>

16. Winans, K.S., Macadam-Somer, I., Kendall, A. et al., 'Life cycle assessment of California unsweetened almond milk', *The International Journal of Life Cycle Assessment*, 10 December 2019, vol. 25, pp. 577–87 <doi.org/10.1007/s11367-019-01716-5>

17. Sultana, M.N., Uddin, M.M., Ridoutt, B.G. et al., 'Comparison of water use in global milk production for different typical farms', *Agricultural Systems*, July 2014, vol. 129, pp. 9–21 <doi.org/10.1016/j.agsy.2014.05.002>

18. Jeffery, C., Schremmer, J. & Condon, M., 'Almond milk might not be as "planet-friendly" as you think, expert sparks fiery debate', *ABC Rural*, 24 February 2021 <abc.net.au/news/rural/2021-02-24/almond-milk-production-and-water-use-fuel-sustainability-debate/13186968>

19. 'Pollinator reliant crops: Almonds', BeeAware <beeaware.org.au/pollination/pollinator-reliant-crops/almonds>

20. Cooper, R., 'Food waste in America: Facts and statistics ', Rubicon, 25 July 2023 <rubicon.com/blog/food-waste-facts>

21. Riddet Institute, 'Feed our Future –"Alternative Proteins – what is their realistic future?"', Vimeo, 18 June 2021 <vimeo.com/564504396>

22. 8.55 billion litres milk produced a year x 3 = ~26 billion litres. 'Australians consuming 2.5 billion litres of drinking milk a year', Seedstock Central, 27 February 2023 <seedstockcentral.com.au/2023/02/27/australians-consuming-2-5-billion-litres-of-drinking-milk-a-year>

23. Silva, B.Q. & Smetana, S., 'Review on milk substitutes from an environmental and nutritional point of view', *Applied Food Research*, June 2022, vol. 2, no. 1, p. e100105 <doi.org/10.1016/j.afres.2022.100105>

24. Mellentin, J., 'Oatly's amazing result: Sales up 12%, losses up 87%', NewNutrition Business, 16 March 2023 <new-nutrition.com/nnbBlog/display/160>

25. Wright, S., 'Trend watch: Is vegan food still on the rise in the UK?', Specialty Food, 28 August 2023 <specialityfoodmagazine.com/retail/trend-watch-is-vegan-food-still-on-the-rise-in-the-uk>

Chapter 16 – Protein Shake

1. Swelum, A.A., El-Saadony, M.T., Abdo, M. et al., 'Nutritional, antimicrobial and medicinal properties of camel's milk: A review', *Saudi Journal of Biological Sciences*, May 2021, vol. 28, no. 5, pp. 3126–36 <doi.org/10.1016/j.sjbs.2021.02.057>

2. Patton, S., 'Some practical implications of the milk mucins', *Journal of Dairy Science*, June 1999, vol. 82 no. 6, pp. 1115–17 <doi.org/10.3168/jds.S0022-0302(99)75334-8>
&
Leischner, C., Egert, S., Burkard, M. et al., 'Potential protective protein components of cow's milk against certain tumor entities', *Nutrients*, 8 June 2021, vol. 13, no. 6 <doi.org/10.3390/nu13061974>

3. Liu, B. & Newburg, D.S., 'Human milk glycoproteins protect infants against human pathogens', *Breastfeeding Medicine*, August 2013, vol. 8, no. 4, pp. 354–62 <doi.org/10.1089/bfm.2013.0016>

4. O'Riordan, N., Kane, M., Joshi, L. et al., 'Structural and functional characteristics of bovine milk protein glycosylation', *Glycobiology*, March 2014, vol. 24, no. 3, pp. 220–36 <doi.org/10.1093/glycob/cwt162>

5. Kaplan, M., Şahutoğlu, A.S., Sarıtaş, S. et al., 'Role of milk glycome in prevention, treatment, and recovery of COVID-19', *Frontiers in Nutrition*, 8 November 2022, vol. 9, article 1033779 <doi.org/10.3389/fnut.2022.1033779>

6. Einerhand, A.W.C., van Loo-Bouwman, C.A., Weiss, G.A. et al., 'Can lactoferrin, a natural mammalian milk protein, assist in the battle against COVID-19?', *Nutrients*, 10 December 2022, vol. 14, no. 24 <doi.org/10.3390/nu14245274>

7. Chutipongtanate, S., Morrow, A.L. & Newburg, D.S., 'Human milk oligosaccharides: Potential applications in COVID-19', *Biomedicines*, 1 February 2022, vol. 10, no. 2 <doi.org/10.3390/biomedicines10020346>

8. Newmark, L.M., 'Milk casein proteins: Ancient, diverse, and essential', *Splash!*, 2018, no. 74 <milkgenomics.org/?splash=milk-casein-proteins-ancient-diverse-essential>

9. Chandran, D., Saleena, L.A.K., Mahesh, S.V. et al., 'Major health effects of casein and whey proteins present in cow milk: A narrative review', *The Indian Veterinary Journal*, November 2021, vol. 98, no. 5, pp. 9–19 <researchgate.net/publication/359025850_Major_health_effects_of_casein_and_whey_proteins_present_in_cow_milk_a_narrative_review>

10. Krissansen, G.W., 'Emerging health properties of whey proteins and their clinical implications', *Journal of the American College of Nutrition*, 1 December 2007, vol. 26, no. 6, pp. 713–23 <doi.org/10.1080/07315724.2007.10719652>

11. Alamiri, F., Riesbeck, K. & Hakansson, A.P., 'HAMLET, a protein complex from human milk, has bactericidal activity and enhances the activity of antibiotics against pathogenic streptococci', *Antimicrobial Agents and Chemotherapy*, 21 November 2019, vol. 63, no. 12 <doi.org/10.1128/aac. 01193-19>

12. Layman, D.K., Lönnerdal, B. & Fernstrom, J.D., 'Applications for α-lactalbumin in human nutrition', *Nutrition Reviews*, June 2018, vol. 76, no. 6, pp. 444–60 <doi.org/10.1093/nutrit/nuy004>

13. Rath, E.M., Cheng, Y.Y., Pinese, M. et al., 'BAMLET kills chemotherapy-resistant mesothelioma cells, holding oleic acid in an activated cytotoxic state', *PLoS One*, 29 August 2018, vol. 13, no. 8, p. e0203003 <doi.org/10.1371/journal.pone.0203003>

14. Sawyer, L., 'β-lactoglobulin and glycodelin: Two sides of the same coin?', *Frontiers in Physiology*, 20 May 2021, vol. 12 <doi.org/10.3389/fphys.2021.678080>

15. Tsutsumi, R. & Tsutsumi, Y.M., 'Peptides and proteins in whey and their benefits for human health', *Austin Journal of Nutrition and Food Sciences*, 2014, vol. 1, no. 1, pp. 1–9 <austinpublishinggroup.com/ nutrition-food-sciences/fulltext/ajnfs-v1-id1002.pdf>

16. Davoodi, S.H., Shahbazi, R., Esmaeili, S. et al., 'Health-related aspects of milk proteins', *Iranian Journal of Pharmaceutical Research*, 2016, vol. 15, no. 3, pp. 573–91 <ncbi.nlm.nih.gov/pmc/articles/ PMC5149046>

17. Córdova-Dávalos, L.E., Jiménez, M. & Salinas, E., 'Glycomacropeptide bioactivity and health: A review highlighting action mechanisms and signaling pathways', *Nutrients*, 12 March 2019, vol. 11, no. 3 <doi.org/10.3390/nu11030598>

18. Kowalczyk, P., Kaczyńska, K., Kleczkowska, P. et al., 'The lactoferrin phenomenon – a miracle molecule', *Molecules*, 4 May 2022, vol. 27, vol. 9 <doi.org/10.3390/molecules27092941>

19. Baier, S.R., Nguyen, C., Xie, F. et al., 'MicroRNAs are absorbed in biologically meaningful amounts from nutritionally relevant doses of cow milk and affect gene expression in peripheral blood mononuclear cells, HEK-293 kidney cell cultures, and mouse livers', *The Journal of Nutrition*, October 2014, vol. 144, no. 10, pp. 1495–1500 <doi.org/10.3945/jn.114.196436>

20. Baier et al., 'MicroRNAs are absorbed in biologically meaningful amounts … '

21. Nguyen, T., 'Unravelling the mysteries of microRNA in breast milk', *Nature*, 17 June 2020, vol. 582, no. 7812 <nature.com/articles/d41586-020-01768-w>

22. Myrzabekova, M., Labeit, S., Niyazova, R. et al., 'Identification of bovine miRNAs with the potential to affect human gene expression', *Frontiers in Genetics*, 11 January 2022, vol. 12, article 705350 <doi.org/ 10.3389/fgene.2021.705350>

23. International Milk Genomics Consortium, '12th international symposium on milk genomics and human health "translation of omics into production and health"', 26–28 October 2015 <milkgenomics. org/wp-content/uploads/2015-IMGC-Symposium-Abstracts.pdf>

24. Sanwlani, R., Fonseka, P., Chitti, S.V. et al., 'Milk-derived extracellular vesicles in inter-organism, cross-species communication and drug delivery', *Proteomes*, 13 May 2020, vol. 8, no. 2 <doi.org/10.3390/ proteomes8020011>

Chapter 17 – You Are What Your Cow Ate

1. den Hartigh, L.J., 'Conjugated linoleic acid effects on cancer, obesity, and atherosclerosis: A review of pre-clinical and human trials with current perspectives', *Nutrients*, 11 February 2019, vol. 11, no. 2 <doi.org/10.3390/nu11020370>

2. Lim, J., Oh, J., Wang, T. et al., 'trans-11 18:1 vaccenic acid (TVA) has a direct anti-carcinogenic effect on MCF-7 human mammary adenocarcinoma cells', *Nutrients*, 10 February 2014, vol. 6, no. 2, pp. 627–36 <doi.org/10.3390/nu6020627>
&
Fan, H., Xia, S., Xiang, J. et al., '*Trans*-vaccenic acid reprograms CD8+ T cells and anti-tumour immunity', *Nature*, 22 November 2023, vol. 623, pp. 1034–43 <doi.org/10.1038/s41586-023-06749-3>
&
Lesté-Lasserre, C., 'Nutrient found in beef and milk shows promise against several cancers', *New Scientist*, 22 November 2023 <newscientist.com/article/2404225-nutrient-found-in-beef-and-milk-shows-promise-against-several-cancers>

3. Lordan, R., Tsoupras, A., Mitra, B. et al., 'Dairy fats and cardiovascular disease: Do we really need to be concerned?', *Foods*, 1 March 2018, vol. 7, no. 3 <doi.org/10.3390/foods7030029>

4. Engel, S., Elhauge, M. & Tholstrup, T., 'Effect of whole milk compared with skimmed milk on fasting blood lipids in healthy adults: A 3-week randomized crossover study', *European Journal of Clinical Nutrition*, February 2018, vol. 72, pp. 249–54 <doi.org/10.1038/s41430-017-0042-5>

5. Średnicka-Tober, D., Barański, M., Seal, C. et al., 'Composition differences between organic and conventional meat: A systematic literature review and meta-analysis', *British Journal of Nutrition*, 16 February 2016, vol. 115, no. 6, pp. 994–1011 <doi.org/10.1017/S0007114515005073>

6. Muller, L. & Delahoy, J., 'Conjugated Linoleic Acid (CLA) in animal production and human health', PennState Extension, 3 January 2023 <extension.psu.edu/conjugated-linoleic-acid-cla-in-animal-production-and-human-health>

7. Butler, G., Nielsen, J.H., Slots, T. et al., 'Fatty acid and fat-soluble antioxidant concentrations in milk from high- and low-input conventional and organic systems: Seasonal variation', *Journal of the Science of Food and Agriculture*, 18 April 2008, vol. 88, no. 8, pp. 1431–41 <doi.org/10.1002/jsfa.3235>

8. Blasko, J., Kubinec, R., Gorova, R. et al., 'Fatty acid composition of summer and winter cows' milk and butter', *Journal of Food and Nutrition Research*, January 2010, vol. 49, no. 4, pp. 169–77 <researchgate.net/publication/236236057_Fatty_acid_composition_of_summer_and_winter_cows%27_milk_and_butter>

9. Muller & Delahoy, 'Conjugated Linoleic Acid (CLA) … '

10. Dei Cas, M., Paroni, R., Signorelli, P. et al., 'Human breast milk as source of sphingolipids for newborns: Comparison with infant formulas and commercial cow's milk', *Journal of Translational Medicine*, 14 December 2020, vol: 18, article 481 <doi.org/10.1186/s12967-020-02641-0>

11. Delgadillo-Puga, C., Cuchillo-Hilario, M., León-Ortiz, L. et al., 'Goats' feeding supplementation with *Acacia farnesiana* pods and their relationship with milk composition: Fatty acids, polyphenols, and antioxidant activity', *Animals*, 1 August 2019, vol. 9, no. 8 <doi.org/10.3390/ani9080515>

12. Soy oil worth US$48 billion in 2021. 'Soybean oil market size, share and COVID-19 impact analysis, by application (cooking and frying, margarine and shortening, salad dressings and mayonnaise, bakery products, and non-food applications), and regional forecast, 2021-2028', *Fortune Business Insights*, January 2022 <fortunebusinessinsights.com/soybean-oil-market-106282>
&
US$59 billion soy meal market in 2021. 'Soybean meal market size, share, competitive landscape and trend analysis report by nature (organic, conventional), by application (food industry, animal feed industry, pharmaceutical), by sales channel (online, offline, b2b): Global opportunity analysis and industry forecast, 2021-2031', Allied Market Research, January 2023 <alliedmarketresearch.com/soybean-meal-market>

13. Soy beans are 20% oil. Yao, Y., You, Q., Duan, G. et al., 'Quantitative trait loci analysis of seed oil content and composition of wild and cultivated soybean', *BMC Plant Biology*, 31 January 2020, vol. 20, article 51 <doi.org/10.1186/s12870-019-2199-7>
&
Soy meal is 2–3% oil. 'Soybean meal', Feedipedia, 4 March 2020 <feedipedia.org/node/674>

14. Cutcliffe, T., 'Omega-3 milk: Grass feeding beats conventional cattle diets', NutraIngredients Europe, 2 March 2018 <nutraingredients.com/Article/2018/03/02/Omega-3-milk-Grass-feeding-beats-conventional-cattle-diets#>

15. Parodi, P.W., 'Cows' milk fat components as potential anticarcinogenic agents', *The Journal of Nutrition*, June 1997, vol. 127, no. 6, pp. 1055–60 <doi.org/10.1093/jn/127.6.1055>

16. Palsdottir, H., '8 Evidence-based health benefits of avocado oil', Healthline, 26 October 2023 <healthline.com/nutrition/9-avocado-oil-benefits#TOC_TITLE_HDR_2>

17. 'Composition of palm oil', M.P. Evans Group PLC, 2023 <mpevans.co.uk/palm-oil/palm-oil-nutrition/composition-of-palm-oil>

18. 'Nutrition: Palm oil and health', Palm Done Right, 21 November 2017 <palmdoneright.com/nutrition>

19. van Vliet, S., Provenza, F.D. & Kronberg, S.L., 'Health-promoting phytonutrients are higher in grass-fed meat and milk', *Frontiers in Sustainable Food Systems*, 1 February 2021, vol. 4, article 555426 <doi.org/10.3389/fsufs.2020.555426>

20. Halin, F., 'Palm oil from the tropics to feed Quebec cows', *Le Journal de Montréal*, 15 February 2021 <journaldemontreal.com/2021/02/15/de-lhuile-de-palme-des-tropiques-pour-nourrir-les-vaches-du-quebec>

21. 'Palm kernel meal, oil 5-20%', Feed Tables <feedtables.com/content/palm-kernel-meal-oil-5-20>

22. Thompson-Morrison, H., Robinson, B. & Gaw, S., 'Feed ingredient for dairy cows could prove harmful', *Modern Farmer,* 7 January 2023 <modernfarmer.com/2023/01/pke-dairy-cows>

23. Salfer, I.J., 'Milk fat above 4% is the new normal', University of Minnesota Extension, 13 April 2022 <extension.umn.edu/dairy-news/milk-fat-above-4-new-normal>

24. Astrup, A., Magkos, F., Bier, D.M. et al., 'Saturated fats and health: A reassessment and proposal for food-based recommendations: *JACC* state-of-the-art review', *Journal of the American College of Cardiology*, 18 August 2020, vol. 76, no. 7, pp. 844–57 <doi.org/10.1016/j.jacc.2020.05.077>

Chapter 18 – Breastmilk Without the Breast

1. Witkowska-Zimny, M. & Kaminska-El-Hassan, E., 'Cells of human breast milk', *Cellular & Molecular Biology Letters*, 13 July 2017, vol. 22, article 11 <doi.org/10.1186/s11658-017-0042-4>

2. Jimenez-Rojo, L., Pagella, P., Harada, H. et al., 'Dental epithelial stem cells as a source for mammary gland regeneration and milk producing cells in vivo', *Cells*, 22 October 2019, vol. 8, no. 10 <doi.org/10.3390/cells8101302>

3. Piñon, N., 'Lab-grown breast milk startup Biomilq aims to change infant nutrition – if it can release a product', CNBC, 25 May 2023 <cnbc.com/2023/05/20/biomilq-startup-makes-lab-grown-breast-milk-amid-baby-formula-issues.html>

4. Rye, M., 'The "green" milk made from cells', BBC Follow the Food <bbc.com/future/bespoke/follow-the-food/the-green-milk-made-from-cells.html>

5. Diamond, R., 'Will cell-based milk change the dairy industry? This California lab could lead the way', Phys.org, 13 October 2022 <phys.org/news/2022-10-cell-based-dairy-industry-california-lab.html>

6. Delisio, E.R., 'Food tech startup bets on the future of cell-based milk', Triple Pundit, 22 February 2022 <triplepundit.com/story/2022/food-tech-cell-based-milk/737486>

7. Riddet Institute, 'Feed our Future –"Alternative Proteins – what is their realistic future?"', Vimeo, 18 June 2021 <vimeo.com/564504396>

8. Fassler, J., 'Lab-grown meat is supposed to be inevitable. The science tells a different story', *The Counter*, 22 September 2021 <thecounter.org/lab-grown-cultivated-meat-cost-at-scale>

9. 'TRANSCRIPT Where's the Beef?', Gastropod, 11 July 2023 <gastropod.com/transcript-wheres-the-beef>

10. Watson, E., 'Cultivated meat: Foodtech fantasy or the future of meat? "None of this stuff makes any commercial sense until everyone's eating it"', AgFunderNews, 1 June 2023 <agfundernews.com/cultivated-meat-foodtech-fantasy-or-the-future-of-meat-none-of-this-stuff-makes-any-commercial-sense-until-everyones-eating-it>

11. Iran Room (FAO Headquarters), 'FAO report on methane emissions in livestock and rice systems: Sources, quantification, mitigation and metrics', Food and Agriculture Organization of the United Nations, 25 September 2023 <fao.org/webcast/home/en/item/6272/icode>

12. Piñon, 'Lab-grown breast milk … '

13. Ruddick, G. & Kollewe, J., 'Unilever to ditch Flora and Stork as consumers turn to butter', *The Guardian*, 6 April 2017 <theguardian.com/business/2017/apr/06/unilever-flora-stork-kraft-heinz-bid>

14. '1992 Margarine outsells butter three to one', Australian Food Timeline <australianfoodtimeline.com.au/margarine-outsells-butter>

15. Randall, G., 'Dairy demand through retail subdued as shoppers scale back', Agriculture and Horticulture Development Board, 22 March 2023 <ahdb.org.uk/news/consumer-insight-dairy-demand-through-retail-subdued-as-shoppers-scale-back>

16. 'Customer case studies: Developing 3D printing technology to create the world's first Cadbury Dairy Milk 3d printer', 3P Innovation <3pinnovation.com/custom-automation/customer-case-studies/developing-3D-printing-technology-to-create-the-world-s-first-cadbury-dairy-milk-3d-printer>

17. Wood, P., 'Cultivated meat review of the cost of manufacturing', *Scribd*, <scribd.com/document/526220188/Cultivated-Meat-review-of-the-cost-of-manufacturing>

18. Stevens, E.E., Patrick, T.E. & Pickler, R., 'A history of infant feeding', *The Journal of Perinatal Education*, 1 January 2009, vol. 18, no. 2, pp. 32–9 <doi.org/10.1624/105812409X426314>

Chapter 19 – The New Moonshine: Raw Milk

1. de Klerk, J.N. & Robinson, P.A., 'Drivers and hazards of consumption of unpasteurised bovine milk and milk products in high-income countries', *PeerJ*, 16 May 2022, vol. 10, p. e13426 <doi.org/10.7717/peerj.13426>

2. 'MARTHA study', MARTHA <martha-studie.de/english-version>

3. Aggeler, M., 'No, you don't need to be drinking raw milk', *The Guardian*, 18 January 2024 <theguardian.com/wellness/2024/jan/17/is-raw-milk-trend-good-bad-for-you-tiktok>

4. Yuan, N., Chi, X., Ye, Q. et al., 'Analysis of volatile organic compounds in milk during heat treatment based on E-nose, E-tongue and HS-SPME-GC-MS', *Foods*, 2 March 2023, vol. 12, no. 5 <doi.org/10.3390/foods12051071>

5. Liang, L., Qi, C., Wang, X. et al., 'Influence of homogenization and thermal processing on the gastrointestinal fate of bovine milk fat: In vitro digestion study', *Journal of Agricultural and Food Chemistry*, 10 November 2017, vol. 65, no. 50, pp. 11109–17 <doi.org/10.1021/acs.jafc.7b04721>

6. Ye, A., Cui, J., Dalgleish, D. et al., 'Effect of homogenization and heat treatment on the behavior of protein and fat globules during gastric digestion of milk', *Journal of Dairy Science*, January 2017, vol. 100, no. 1, pp. 36–47 <doi.org/10.3168/jds.2016-11764>

7. van Lieshout, G.A.A., Lambers, T.T., Bragt, M.C.E. et al., 'How processing may affect milk protein digestion and overall physiological outcomes: A systematic review', *Critical Reviews in Food Science and Nutrition*, 5 August 2020, vol. 60, no. 14, pp. 2422–45 <doi.org/10.1080/10408398.2019.1646703>

8. Liang, N., Mohamed, H.M., Kim, B.J. et al., 'High-pressure processing of human milk: A balance between microbial inactivation and bioactive protein preservation', *The Journal of Nutrition*, September 2023, vol. 153 no. 9, pp. 2598–611 <doi.org/10.1016/j.tjnut.2023.07.001>

9. Langer, A.J., Ayers, T., Grass, J. et al., 'Nonpasteurized dairy products, disease outbreaks, and State laws – United States, 1993–2006', *Emerging Infectious Diseases*, March 2012, vol. 18, no. 3 <doi.org/10.3201/eid1803.111370>

10. '*QuickStats*: Percentage of adults aged ≥20 years who consumed dairy on a given day, by amount and sex – National Health and Nutrition Examination Survey, United States, 2011–2012', Centers for Disease Control and Prevention, 17 July 2015 <cdc.gov/mmwr/preview/mmwrhtml/mm6427a5.htm>

11. Ogunwole, S.U., Rabe, M.A., Roberts, A.W. et al., 'U.S. adult population grew faster than nation's total population from 2010 to 2020', United States Census Bureau, 12 August 2021 <census.gov/library/stories/2021/08/united-states-adult-population-grew-faster-than-nations-total-population-from-2010-to-2020.html>

12. Lando, A.M., Bazaco, M.C., Parker, C.C. et al., 'Characteristics of U.S. consumers reporting past year intake of raw (unpasteurized) milk: Results from the 2016 Food Safety Survey and 2019 Food Safety and Nutrition Survey, *Journal of Food Protection*, July 2022, vol. 85, no. 7, pp. 1036–43 <doi.org/10.4315/JFP-21-407>

13. Koski, L., Kisselburgh, H., Landsman, L. et al., 'Foodborne illness outbreaks linked to unpasteurised milk and relationship to changes in state laws – United States, 1998–2018', *Epidemiology & Infection*, 2022, vol. 150, p. e183 <doi.org/10.1017/S0950268822001649>

14. 'Contribution of different food commodities (categories) to estimated domestically-acquired illnesses and deaths, 1998-2008', Centers for Disease Control and Prevention, 29 January 2013 <cdc.gov/foodborneburden/attribution-image.html#foodborne-illnesses>
&
Bennett, S.D., Sodha, S.V., Ayers, T.L. et al., 'Produce-associated foodborne disease outbreaks, USA, 1998–2013', *Epidemiology & Infection*, August 2018, vol. 146, no. 11, pp. 1397–1406 <doi.org/10.1017/S0950268818001620>

15. Bennett et al., 'Produce-associated foodborne disease outbreaks … '

16. 'Large haul of bathtub cheese in Riverside County', California Department of Food and Agriculture, 27 January 2006 <cdfa.ca.gov/egov/Press_Releases/Press_Release.asp?PRnum=06-004>

17. Piatt, R., '"Bathtub cheese" linked to 2,000 cases of Salmonella', KSL.com, 3 November 2011 <ksl.com/article/17939543/bathtub-cheese-linked-to-2000-cases-of-salmonella>

18. Willoughby, K., 'The evolution of milk pasteurisation', *Dairy Global*, 5 July 2022 <dairyglobal.net/dairy/milking/the-evolution-of-pasteurisation>

Chapter 20 – Hog Slop

1. 'History of low-fat milk', *Food History*, 25 April 2023 <world-foodhistory.com/2023/04/history-of-low-fat-milk.html>

2. Smith-Howard, K., *Pure and Modern Milk: An environmental history since 1900*, Oxford: Oxford University Press, 2013

3. Smith-Howard, K., 'Would you wear a sweater made from milk?', *Slate*, 2 February 2014 <slate.com/technology/2014/02/uses-for-skim-milk-before-it-was-marketed-as-a-nonfat-diet-product-hog-slop-and-wool.html>

4. Smith-Howard, K., *Pure and Modern Milk: An Environmental History Since 1900*, New York: Oxford University Press, 2014
5. Liang, L., Qi, C., Wang, X. et al., 'Influence of homogenization and thermal processing on the gastrointestinal fate of bovine milk fat: In vitro digestion study', *Journal of Agricultural and Food Chemistry*, 10 November 2017, vol. 65, no. 50, pp. 11109–17 <doi.org/10.1021/acs.jafc.7b04721>
6. 'Innovative proteins', Ingredia <ingredia.com/fields-of-expertise/functional-nutritional-proteins/innovative-proteins>

Chapter 21 – Blessed They Are

1. Alberding, F., 'Pig's milk cheese is tasty, but it won't make you rich', Munchies Food by Vice, 25 August 2015 <vice.com/en/article/8qkkmb/pigs-milk-cheese-is-tasty-but-it-wont-make-you-rich>

Chapter 22 – Answering the Big Questions

1. Bartley, J. & McGlashan, S.R., 'Does milk increase mucus production?', *Medical Hypotheses*, April 2010, vol. 74, no. 4, pp. 732–4 <doi.org/10.1016/j.mehy.2009.10.044>
2. Lukito, W., Malik, S.G., Surono, I.S. et al., 'From "lactose intolerance" to "lactose nutrition"', *Asia Pacific Journal of Clinical Nutrition*, December 2015, vol. 24, no. 1, pp. 1–8 <doi.org/10.6133/apjcn.2015.24.s1.01>
3. Argnani, F., Di Camillo, M., Marinaro, V. et al., 'Hydrogen breath test for the diagnosis of lactose intolerance, is the routine sugar load the best one?', *World Journal of Gastroenterology*, 28 October 2008, vol. 14, no. 40, pp. 6204–7 <doi.org/10.3748/wjg.14.6204>
4. Reed, K.E., Camargo, J., Hamilton-Reeves, J. et al., 'Neither soy nor isoflavone intake affects male reproductive hormones: An expanded and updated meta-analysis of clinical studies', *Reproductive Toxicology*, 1 March 2021, vol. 100, pp. 60–7 <doi.org/10.1016/j.reprotox.2020.12.019>

Index